SOTHIS PRESS

Sothis Press aims to completely reimagine the public domain book with an emphasis on universal access to captivatingly designed print editions. We believe that technology will enable every single book ever published to be perpetually accessible in beautiful print form. Sothis is beauty in ubiquity.

www.sothis.press
1 Waterhouse Square, London, ECIN 2ST

Small Boat Navigation

Frank W. Sterling

SOTHIS PRESS

ISBN 978-1-0368-0673-6

This edition is published by
Sothis Press © 2023. The company
is registered in England and Wales.
The Company Number is 14900389.

CONTENTS

PART I

NAVIGATION

PART II

SEAMANSHIP

PART I

NAVIGATION

SMALL BOAT NAVIGATION

CHAPTER I

NAVIGATION

NAVIGATION is that science which affords the knowledge necessary to conduct a vessel from one point to another on the earth and provides a means of determining the position of the vessel at any time. There are three general branches to this science, viz:

(1) Piloting.

(2) Dead Reckoning.

(3) Nautical Astronomy.

Piloting is determining the position of a vessel by observations on known visible charted objects, or by soundings of the depth of the sea.

Dead Reckoning, or the method of sailings, is a means of deducing a vessel's position from the direction and distance sailed from a previous known position. This method involves the rules of plane trigonometry.

Nautical Astronomy treats of the determination of a vessel's position by observation of celestial bodies — the sun, stars, moon, and planets. It is based on spherical trigonometry and its use is principally confined to deep sea navigation. On this account it is not dealt with in this work.

UNITS DEFINED

The Statute Mile, which is 5,280 feet, is employed in land measurements. It is used to some extent in navigating river and lake vessels, especially on the Great Lakes, but by far the more commonly used unit is

The Geographical or Nautical Mile. This unit of linear measurement, used by navigators, is generally termed the nautical or *sea mile.* It is approximately 6,080 feet. It has various values as defined by the standards of different countries, but from the navigator's standpoint these various standards do not vary materially from the above value. It is equal to a *minute of arc of the equator.* For purposes of navigation the nautical mile is assumed equal to a minute of latitude at all points of the earth. Hence, when a vessel changes her position to the north or south by one nautical mile, it is assumed that the latitude has changed 1′. Owing to the fact that the meridians converge toward the poles, the difference in longitude due to a change of position of one mile to the east or west varies with the latitude.

Whereas a *departure* (change of position to east or west) of 1 mile at the equator equals 1′ of arc, the same departure of 1 mile in a latitude of 60° amounts to 2′ in longitude.

The *Knot* is the measure of speed and equals *one nautical mile per hour.*

The Axis of the Earth is a diameter passing through the poles of the earth and about which the earth revolves.

The Terrestrial Equator is a great circle of the earth passing through the middle point of this axis and perpendicular thereto. It divides the earth into the Northern and Southern hemispheres, and every point on this equator is equidistant from the poles. Longitude is measured along the equator.

Terrestrial Meridians are great circles of the earth passing through the poles. They are perpendicular to the equator. Latitude is measured on the meridians, being 0° at the equator and 90° at the poles. The *Meridian* of a place is that meridian passing through the place.

The Prime Meridian is that meridian from which longitude is measured. The meridian passing through Greenwich, England, is almost universally accepted as the prime meridian for navigation. Longitude is measured from 0° at Greenwich, East and West to 180°.

Parallels of Latitude are small circles of the earth parallel to the equator.

The Latitude of a place is the arc of the meridian intercepted between the equator and that place. It is reckoned north and south from the equator as an origin, up to 90° at the poles.

The Longitude of a place is the arc of the equator intercepted between the meridian of the place and the prime meridian. Longitude is measured East and West through 180° from the meridian of Greenwich.

The Difference of Latitude of two places is the portion of a meridian included between the two parallels of latitude passing through the two places. When two places are on the same side of the equator the *difference of latitude* is the numerical difference between the latitudes of the places; when on opposite sides of the equator the *difference of latitude* is the numerical sum of the two places. The difference of latitude is called North or South to indicate in which direction a vessel would sail to make the change.

The Difference of Longitude of two places is the arc on the equator intercepted between the meridians passing through the two places. When the places are in the same longitude (viz: East or West), *the difference of longitude* is the numerical difference between the longitudes of the two places; when in different longitudes, ***the difference in longitude*** is the numerical sum of the two longitudes, or 360° minus this sum. The difference of longitude is marked East or West to denote in which direction a vessel would sail to make the change.

CHAPTER II

NAVIGATIONAL INSTRUMENTS, BOOKS, RECORDS, ETC.

1. NAVIGATIONAL INSTRUMENTS

THE following instruments are indispensable to the navigator when piloting: (a) dividers or compasses, (b) parallel rulers, (c) log, chip or patent, (d) lead and line, (e) compass, liquid, dry, or gyroscopic, (f) azimuth circle, (g) barometer, mercurial or aneroid, (h) thermometer. A sextant and protractor may also be included in the outfit, although the former is entirely unnecessary.

(a) DIVIDERS (OR COMPASSES) consist of two pointed legs movable about a joint so that the points at the extremities of the legs may be set at any desired distance apart. *Dividers* is an instrument used for measuring distances on a chart. One of the pointed extremities can be replaced by a pencil or pen for describing arcs or circles and in this case the instrument is called a *compass*.

(b) PARALLEL RULERS consist of two wooden straight edges, joined by two metal links, so constructured that in all positions of the links the straight edges are always parallel to each other. It is used to draw lines parallel to each other, and especially in chart work for transferring lines on the chart to the compass rose, for laying off courses by transferring them from the compass rose, and for plotting bearings of objects.

(c) The Log is an instrument for measuring or estimating the speed of the vessel and the distance run for any given period. It takes a variety of forms that may be classified under two heads, the chip log (which measures the speed of the vessel at any instant) and the patent log (which is cumulative and measures the distance run for any interval).

The Chip Log, which is quite inexpensive and can be made by the navigator, consists of three parts, the *log-chip*, the *log-line*, and the *log-glass*. Its principle is that, if a light object is thrown from the ship, it ceases to partake of the motion of the ship as soon as it strikes the water. If after any known interval of time its distance from the ship is known (this being equivalent to the distance the ship has travelled in the known interval of time) the approximate speed of the ship can be computed.

The *log-chip* is a thin wooden quadrant of about 5″ radius, the circular edge being weighted with lead so that the chip will float upright, apex up. The log-line is knotted into a hole at the apex. The ends of a bight of line are knotted into the other two corners of the chip. Into this bight is seized a wooden peg which fits into a

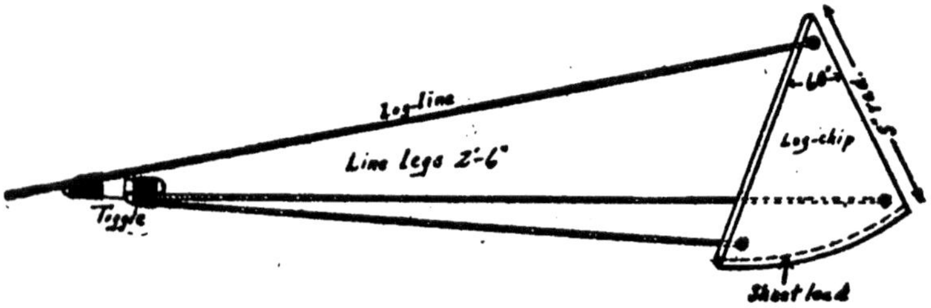

Fig. 1.—Chip Log.

socket seized on the log-line. When the peg is in the socket, all three lines to the corners of the chip are of equal length. This contrivance facilitates hauling in the chip after the speed is obtained.

The *log-line,* which is made of halyard stuff, is about 150 fathoms long. One end is secured to the chip and the other to a reel on which the line is wound. The line is marked at about 15 fathoms from the chip end by a piece of bunting. This part of the line is called *stray line.* From this piece of bunting the line is marked at every 47 feet 3 inches by a piece of fish line held between the strands of the log-line, the line being marked by a knot in the fish line for every division (47 feet 3 inches) from the bunting. Thus at 94 feet

6 inches from the bunting the piece of fish line has two knots in it, etc. These main divisions, called knots, are further subdivided into five equal parts by pieces of white bunting between the strands to indicate two-tenths of a knot.

The *log-glass* is a sand glass similar to an hour glass constructed to run for 28 seconds. A 14-second glass is also used.

Three men are needed to "heave the log." One heaves the chip-log and tends the log-line, one holds the reel, and one tends the log-glass.

To find the speed by the chip-log, hold the reel by its handles and unwind some of the stray line. Insert the toggle in its socket and heave the chip overboard, allowing the line to run out freely. As the first piece of bunting, which marks the end of the stray line, passes over the rail invert the log-glass. Just as the last particle of sand passes from the top to the bottom of the glass seize the log-line, which has been running out freely. The subdivisional mark which is now at the rail indicates the speed of the vessel in knots and tenths. For instance, if the cord having three knots is at the rail, the vessel is making three knots per hour. This can be demonstrated as follows:

Principle of Construction. When the chip hits the water it ceases to partake of the motion of the ship and becomes stationary in the water.

Between the first mark and the third the interval of time is 28 seconds (the time it takes the sand to run from the top to the bottom of the glass). In this interval of time the vessel moves 3 times 47 feet 3 inches (as shown by the log-line). Now in feet $\frac{3 \times 47.25 \times 60 \times 60}{28}$ is the distance that the vessel would move in one hour at the same rate.

Or $\frac{3 \times 47.25 \times 60 \times 60}{28 \times 6080} = 3$ knots per hour.

The 28-second glass is used for low speeds. For speeds over 5 knots a 14-second glass is used and the reading of the log-line is doubled.

To haul in the line after a reading is obtained, give the line a sharp tug. This will release the toggle and the chip will lay flat on the surface and can be hauled in hand over hand and reeled up.

The whole apparatus should be overhauled frequently to check the log-line markings and the log-glass. The line must be frequently checked and remarked as it stretches or shrinks, and should be marked when wet. The glass is checked by comparing it with a watch, and the sand is dried if it becomes damp as indicated by requiring more than 28 seconds to pass from the top to the bottom of the glass.

Figure 1 shows the parts of the chip-log and

the necessary dimensions for making one of these instruments.

In lieu of the *log glass* furnished with commercial outfits time may be kept with the watch and for convenience a 30-second interval may be substituted for the 28-second interval indicated by the log glass. In this case the knot marks on the log line are placed 50 feet 8 inches apart.

The principle of marking the log line can be checked easily. Suppose the vessel is moving at the rate of one knot (6080 feet per hour). Then in 30 seconds it would move 50 feet 8 inches. This is the length between markings. If the line is accurately marked and the log properly constructed the speed indicated by the chip log will be one knot.

The Patent *Log* is a mechanical contrivance for measuring the actual distance that a vessel travels through the water. It is sometimes called a taffrail log because it is usually installed on the taffrail. It takes a variety of forms, but all are more or less subject to inaccuracies and all embody the same principles.

The patent log consists of three parts, (1) the rotator, a conical shaped piece of brass fitted with vanes which is towed astern and caused to rotate as it passes through the water at a speed proportionate to the speed of the vessel; (2) a register, located on the vessel's rail. This register is connected to the rotator by a line which communi-

FIG. 2.—Negus Taffrail Log.

cates the revolutions of the rotator to cyclometer gear in the register. The whole is so calibrated that the miles and tenths run by the vessel are registered by appropriate dials on the register; (3) the line, which is specially made. Usually about 400 feet of line is used to connect the rotator and the register. A high speed requires a longer line than a low speed. The parts of the Negus Taffrail Log are shown in Figure 2.

Patent logs are not exact instruments and many inaccuracies enter even when in good working order. They must be carefully watched. If correct at one speed they may be inaccurate at a faster or slower speed; they register differently in a head and a following sea, and in smooth and rough weather. The error of the patent log should be determined for varying conditions of sea and at different speeds for every run between two ports, if the speed is not affected by tide, and a record of errors under these varying conditions should be kept to correct future readings.

The revolutions of the screw propeller afford the most accurate measure of the speed of a vessel and the distance steamed. The revolutions of the propeller (engine speed) can be obtained by a small instrument called a tachometer. By running over a measured mile (or other known distance) at various engine speeds, the revolutions corresponding to various speeds can be obtained and tabulated thus:

Revolutions	*Speed in Knots*
85	7
97	8
110	9
125	10
142	11
160	12 etc.

Entering this table with the average revolutions for any interval of time, the corresponding speed can be obtained. While this is the most accurate method of getting the speed, it must be borne in mind that by whatever of the above methods the speed is obtained, it is the *speed through the water* and that to obtain the *speed made good over the ground* allowances must be made for the local current.

(d) THE LEAD AND LINE is a device for ascertaining the depth of water. It consists of a suitably marked line, having a long hexagonal or octagonal lead at its end. Two sizes of leads are used, one weighing from 7 to 14 pounds, called the *hand lead*, and used for depths up to 25 fathoms, and the other weighing from 30 to 100 pounds, called the *deep sea lead*, and used for depths over 25 fathoms.

Lead lines are marked, measuring from the bottom of the lead secured to the line, as follows:

At 2 fathoms with two leather strips,
3 fathoms with three leather strips,
5 fathoms with a white rag,
7 fathoms with a red rag,
10 fathoms with a leather having a hole in it,
13 fathoms the same as 3,
15 fathoms the same as 5,
17 fathoms the same as 7,
20 fathoms with two knots,
25 fathoms with one knot,
30 fathoms with three knots,
35 fathoms with one knot, etc.

Sometimes the lead is marked in feet around the depth corresponding to the vessel's draft. Soundings that correspond to depths marked on the lead line are called "*marks*"; intermediate soundings are called "*deeps*." Lead lines should be marked when wet, and should frequently be verified and remarked when necessary. The bottom of the deep sea lead is hollow. When taking a sounding this cavity is filled with tallow (called "arming the lead") which picks up a sample of the bottom. This allows a comparison with the character of the bottom as indicated on the chart.

THE SOUNDING MACHINE is an instrument for obtaining rapid soundings at great depth. It is employed on all large ships and is described at length in Chapter 5.

(e) THE COMPASS may take one of three forms, liquid, dry, or gyroscopic. The liquid compass is the one most commonly used in this country and will be described later. The dry compass is used extensively in England. The gyroscopic compass is used in the U. S. Navy; its high cost precludes its use in any but the largest ships.

The Wet Compass consists of a skeleton card 7½″ or less in diameter, made of tinned brass, resting on a pivot in liquid, with provision for two pairs of magnets symmetrically placed. The magnets consist of four cylindrical bundles of steel wire that are magnetized as bundles between the poles of an electromagnet. They are cased in cylinders and secured to the underside of the compass card in a direction parallel to the North and South markings of the card.

The card is curved, annular shape or flat, graduated by quarter points, and a card circle is adjusted to its outer edge graduated to half degrees, with readings at every point and every five degrees, as seen in Figure 3.

The card is secured on a concentric spheroidal air vessel which rests on liquid in the compass bowl. The air vessel is fitted with an agate bearing which rests on a pivot in the compass bowl, most of the weight of the card being supported by the liquid and only a slight pressure being on the pivot.

The compass bowl of cast bronze has a glass cover so that the card can be seen. The cover is made tight by a rubber gasket. The liquid in the bowl is 45% pure alcohol and 55% distilled water. The inside of the bowl is painted white and a lubber's point is drawn in the bowl in the

FIG. 3.—Compass Points.

fore and aft line of the ship. This is the reference point when reading the compass.

The under side of the bowl constitutes an expansion chamber which allows for expansion with a change of temperature of the liquid in the bowl.

The bowl is mounted by double gymbals on a composition stand. This stand contains magnets to *compensate the compass.* The bowl can be covered by a spun brass hood which fits on the stand and can be revolved for taking bearings.

Boxing the Compass is the process of naming the points in their order. The four points,

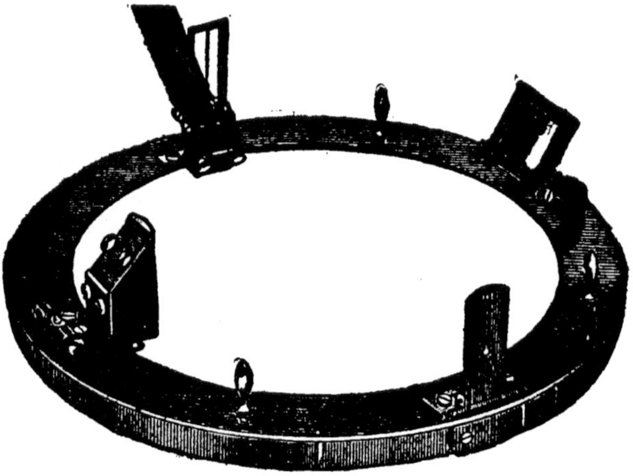

FIG. 4.— Azimuth Circle.

North, East, South, and West, are called *cardinal* points. Midway between the cardinal points are the *intercardinal* points, N.E., S.E., S.W., N.W. The names of all points are shown in Figure 3.

(f) THE AZIMUTH CIRCLE consists of a composition ring which fits over the edge of the compass bowl, Figure 4. At one extremity of a diameter is a curved mirror hinged to move about a horizontal axis, and facing this, at the other extremity of the diameter, is a cased prism which has a narrow slit in the case facing the mirror. This part of the instrument is used to take azimuths of the sun for compass work.

At the extremities of another diameter (at right angles to the first) is a hinged vertical wire, and a hinged plate having a vertical slit. This is used for taking direct bearings of an object. Hinged on the same pivot as the vertical wire is a smoked glass reflector for azimuth work with the sun or stars.

A level on the azimuth rim shows when the circle is horizontal. All observations must be taken with a horizontal azimuth circle.

(g) THE BAROMETER is an instrument for measuring the atmospheric pressure. Barometric observations are necessary for weather predictions. Some form of barometer is necessary on all vessels. There are two forms, the Mercurial Barometer, and the Aneroid Barometer. The mercurial barometer is carried on large yachts. Smaller boats should be equipped with the aneroid.

The Mercurial Barometer indicates atmospheric pressure by the height of a column of mercury. If a glass tube, closed at one end, is entirely filled with mercury and then placed open end down over a cup of mercury (no mercury being allowed to escape from the tube during the operation) the mercury in the tube will fall until its level is about 30 inches above the level of the mercury in the cup. This column of mercury, in the tube, is sustained by the pressure of air on the mercury in the cup, and will rise or fall with this pressure (which is atmospheric pressure).

This is in effect a barometer. In practice the cup and tube are encased in a brass case, cut away near the level of mercury in the tube; along this opening is a scale for reading the height of mercury in inches, a *vernier* being fitted for accurate readings.

The Vernier is an attachment used on many instruments, such as barometers, sextants, protractors, etc., to facilitate exact readings. It consists of a metal scale, similar in construction to that of the scale to which it is fitted and arranged to move along the main scale by a rack and pinion.

The vernier scale has a total length equal to one of the whole divisions of the main scale, but this length is divided into one more or one less part than the number of subdivisions into which the whole division of the main scale is divided.

Suppose that a barometer scale is divided into tenths of an inch and that a length of nine such divisions be divided into ten parts for a vernier, Figure 5. Number the vernier divisions from 0 at the bottom to 10 at the top. If the bottom division of the vernier is brought level with the top of the mercurial column the scale is read as follows: In Figure 5 the mercury stands above the mark 29.6 on the main scale. Find the division of the vernier that coincides with a division of the main scale. In the figure this division is "1." Therefore .01 must be added and the exact reading is 29.61.

The Aneroid Barometer is an instrument by which the atmospheric pressure is measured by the elasticity of a metal plate. It consists of a cylindrical brass box, the metal being very thin. This box is in a state of partial vacuum. As the atmospheric pressure increases the enclosed air is compressed and the ends of the box approach each other. Suitable levers communicate this motion of the box ends to a pointer that moves over a suitable scale. These levers magnify the motion of the box ends and the scale is calibrated to indicate the air pressure.

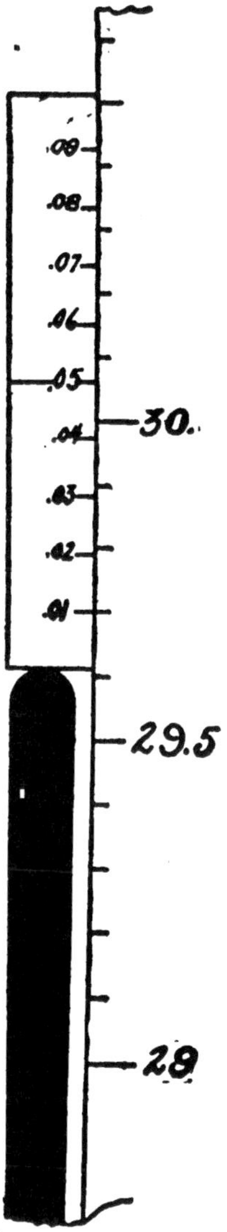

FIG. 5.— Vernier Scale.

Barometer Comparisons. Whatever the form of barometer used, it will be subject to errors due to derangements or to inaccurate construction. Because of this it is necessary to compare the barometer frequently with a standard of known error. At all principal ports a standard mercurial barometer is available for comparison. From this comparison the barometer error can be computed and applied to future readings. At the principal ports the reading of a standard barometer at specified times is published in the daily papers. By observing the barometer on board at the same time the error of the vessel's instrument is obtained.

(h) THE THERMOMETER is an instrument for measuring temperatures. Its principle is too well known to require discussion. It is an aid to the mariner in predicting weather, judging the humidity of the atmosphere, and, through the temperature of the sea water, finding proximity of currents such as the Gulf Stream. Sea water used for observations should be drawn from at least three feet below the surface.

THE PSYCHROMETER consists of two thermometers called the *wet and dry bulbs*. The dry-bulb thermometer gives the temperature of the air. The wet-bulb thermometer is exactly like the dry-bulb except that its mercurial bulb is surrounded with cloth which is kept moist. It indicates the *temperature of evaporation*. By reading both bulbs the humidity of the air is obtained and probability of rain can be foretold.

THE SEXTANT[1] is one of the most valuable of deep sea navigational instruments, but is of little

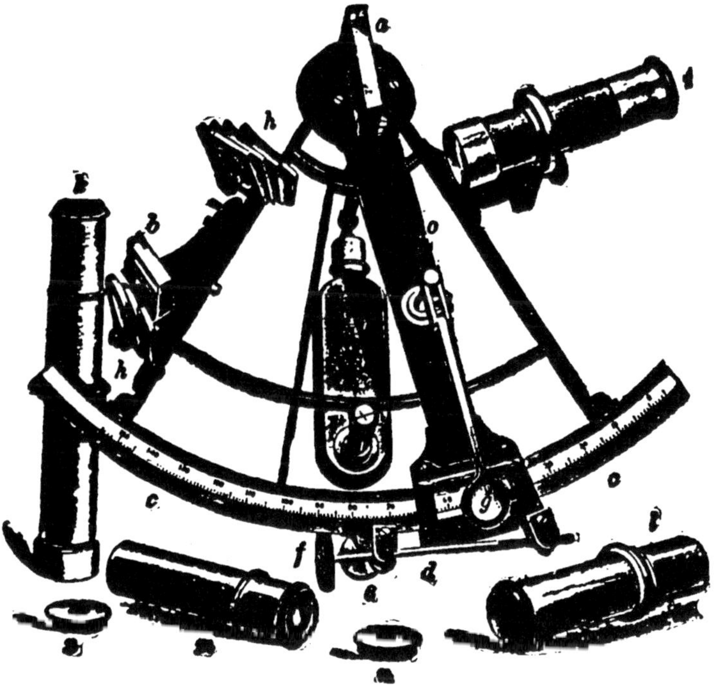

FIG. 6.— Sextant.

use when coasting. It is used to obtain the angle between two objects, terrestrial or celestial, by bringing into coincidence at the observer's eye the rays of light from the two objects, one ray direct

[1] Although the sextant is of little use for motor boats its description and method of adjustment is given here for general information.

and the other ray by double reflection. Its general form is shown in Figure 6. The frame is of brass or other composition. The graduated arc (cc) is generally of silver, the graduations being by degrees and minutes, the smallest subdivision representing generally 10′. By means of the vernier (d) the instrument can be read to smaller divisions, usually 10″, 15″, or 20″. The magnifying glass (g) facilitates reading the vernier. A wooden handle is fitted for holding the instrument. A brass index arm (o) carrying the vernier (d) is pivoted at the center of the arc (cc). It carries the index glass (a) which moves with the pivot of the index arm (o) as an axis. A second glass (b), called the horizon glass, is fixed to the frame. It is half mirror and half transparent, the line of demarcation being parallel to the plane of the instrument. Both glasses are perpendicular to the plane of the instrument and are provided with adjusting screws to permit of adjustment.

The index arm can be clamped in position on the arc (cc) by the clamp (e) and the tangent screw (f) can give the arm small movements after it is clamped. A telescope (i), supported in an adjustable ring on the frame, is used to give greater distinction to the images. In lieu of this telescope the star telescope (k) or plain sighting tube

(l) may be used. Colored shades (hh) are fitted before the index and horizon glasses to protect the eye from sun glare. The same result obtains by using the colored cap (n) on the telescope.

The vernier (d) is constructed on the same principle as that explained under the barometer.

To Use the Sextant for measuring angles proceed as follows: Point the telescope at the lower object, if one is above the other, or at the left hand object, if both are in nearly the same horizontal plane. Keep this object in direct view through the transparent part of the horizon glass, and move the index arm until the reflection of the second object is seen in the silvered part of the horizon glass. When the objects are nearly in coincidence, clamp the index arm and by the tangent screw bring the objects in exact coincidence at the line of demarcation on the horizon glass. The angle between the objects is now shown on the arc. When measuring the height of an object above the horizon the nearest point of the horizon, directly below the object must be employed. To find this point swing the instrument about the line of sight as a center, keeping the image of the object in the middle of the field. The object will appear to describe the arc of a circle, the lowest point of which marks the vertical. When bringing a celestial object down to the horizon it is usual to set the instrument to zero and point it at the object. Then keeping the object in the field move the index arm until the horizon appears in the field.

Adjustments of Sextant. The adjustments ordinarily made by the navigator are to keep the horizon and index glasses perpendicular to the plane of the instrument, and the line of sight parallel to the plane of the instrument.

Adjustment of the Index Mirror. Clamp the arm near the middle of the arc. Place the eye near the index mirror and sight along the plane of the instrument. If the direct and reflected images of the arc appear in one plane the index mirror is in adjustment, perpendicular to the plane of the instrument. If the reflected image of the arc appears to droop from the direct image the glass leans backward; if it seems to rise the glass leans forward. Adjust by screws back of the mirror.

Adjustment of the Horizon Mirror. Having adjusted the index mirror, select a celestial object, preferably a star, to adjust the horizon mirror. Put in the telescope and direct it toward a star. Move the index arm until the reflected image passes the direct image. If one passes directly over the other the mirror is in adjustment. If one passes to one side of the other, adjust the horizon mirror by its attached screws until the images pass one over the other.

Adjustment of the Line of Sight is more difficult. Screw the star telescope, which has two parallel wires in its eye piece, into the ring; turn the eye piece until the wires are parallel to the plane of the instrument. Select two well defined objects whose angle is greater than 90° (stars preferred). Bring these objects into coincidence at one wire of the eye piece. Now move the instrument until they are seen at the other wire. If still in coincidence the line of sight is in adjustment. If not, correct by the adjusting screws of the ring.

The Index Error of a sextant is an error in its reading due to the fact that when both mirrors are parallel the zero of the vernier does not coincide with the zero of the arc. This error does not necessarily remain constant and it is good practice to determine it each time the instrument is used.

The Index Correction, which is the index error with its algebraic sign reversed, is the correction that must be applied to an observed angle to get the true angle. It may be found by observation on (a) a star, (b) the sea horizon, (c) the sun.

(a) Bring the direct and reflected images of the star into coincidence and read the sextant.

This reading is the index correction, and is + or — according as the vernier zero is to the right or left of the zero on the arc.

(b) Proceed in a similar manner, substituting the sea horizon for the star.

(c) Bring the upper limb of the direct image of the sun tangent to the reflected lower limb. Read the instrument and mark the reading + or — as in (a). Now bring the lower limb of the direct image tangent to the reflected upper limb. Read the instrument and give the proper algebraic sign according to the above rule. *One half the algebraic sum of the two readings is the index correction.*

Of these three methods the first is preferable as it is the most accurate. Always make contact by moving the tangent screw in the same direction.

THE KOCH PROTRACTOR is one of the most useful instruments known to piloting. It combines all the facilities of a portable compass rose, the parallel rulers, the protractor, and the plotter. It is described at length in the next chapter.

2. BOOKS AND ACCESSORIES

(a) CHARTS. A full set of coast and harbor charts should be on hand for all waters that it is intended to visit, or that any emergency might cause the owner to visit. *Before sailing care* must be taken to see that these charts are corrected to date. A weekly bulletin issued by the U. S. Hydrographic Office, Washington, D. C., supplies all corrections and can be had on request. An uncorrected chart is not infrequently the cause of grounding. Large scale harbor charts should be obtained for all harbors and inland waters that may be visited.

The following various publications, issued by the Hydrographic Office, dealing with special features of navigation, should be regularly consulted. They can be found at any branch Hydrographic Office in the large sea ports.

Pilot Charts of the various oceans, which furnish information about drifting derelicts, ice and floating obstructions, storm tracks, average wind and weather, ocean currents, etc.

Hydrographic Bulletin, weekly, which gives weekly changes of the above.

Daily Memorandum, which gives daily information of interest to mariners.

Notices to Mariners, which are weekly bulletins of changes in aids to navigation (lights, buoys, etc.), dangers to navigation (rocks, shoals, bars, etc.), and in general all facts that affect charts, sailing directions, etc. All charts and sailing directions should be kept corrected to date from these notices.

(b) A Light and Buoy List of latest date must be carried. It consists of a list of lights and buoys with their location (Latitude and Longitude), their descriptions and characteristics, and other information such as fog signal stations and submarine signal stations. It can be obtained from the Hydrographic Office and must be kept corrected to date in a similar manner to, and from the same source as, the charts.

(c) Sailing Directions of that part of the world to be navigated should be carried. They come in bound volumes, each volume covering a large tract of pilot waters. They give detailed information of harbors, coasts, currents, courses for entering harbors, cable, provision and coaling facilities, in fact are indispensable fonts of navigational data. They are obtained from the same source as the charts and are corrected from the same publications. For American waters use U. S. Coast Pilots.

(d) Bowditch's Useful Tables (latest date) should be used. This book contains 48 different tables, and these comprise all the tables needed for any coasting or piloting problem that may arise.

(e) Tide Tables and the Nautical Ephemeris of the current year may be carried but are not necessary while on pilot waters. The Nautical

Ephemeris will be needed if any celestial observations are made, and in this case a well rated chronometer will also be necessary. The time of high and low water for any port may be obtained from the local paper.

3. RECORDS THAT SHOULD BE KEPT

(a) THE LOG BOOK is a record of the vessel's cruise, and is a most necessary accessory. It should contain all the data of the navigation by dead reckoning, and should afford a complete meteorological record. In addition all occurrences of note should be recorded. It is the only available legal record in case of crime or accident on board.

Hourly data. The following hourly data should be entered in the log at the *end* of each hour.

1. Knots, to nearest tenths, made good during the hour.
2. Patent log reading, if one is carried.
3. The average engine revolutions for the hour.
4. Courses steered during the hour. (The exact time of changing course, with the patent log reading at that time should be recorded. This data is used to work up the dead reckoning, which is discussed later.)
5. The wind, its force and direction.
6. The barometer, and its attached thermometer.
7. The thermometer, both wet and dry bulb.
8. The temperature of the sea water.
9. State of weather.
10. Clouds, form, quantity, and direction from which moving.
11. State of sea.

The information in 1, 2, 3, and 4 is used for navigation.

The information in 5, 6, 7, 9, 10, and 11 is used to foretell the weather. (8) The temperature of the sea water will help detect the proximity of ice, or the presence of a cold or hot current, such as the Gulf Stream.

The Force of the Wind (5) is recorded numerically from 0 (a calm) to 12 (a hurricane). Admiral Beaufort devised a convenient scale which is given in part below:

Force of Wind.	*Velocity in Statute Miles per Hour.*
0 — Calm	0–3
1 — Light air	8
2 — Light breeze	13
3 — Gentle breeze	18
4 — Moderate breeze	23
5 — Fresh breeze	28
6 — Strong breeze	34
7 — Moderate gale	40
8 — Fresh gale	48
9 — Strong gale	56
10 — Whole gale	65
11 — Storm	75
12 — Hurricane	90 and over.

The State of the Weather (9) is entered in the log by symbols as follows:

b — Clear blue sky.
c — Clouds present in sky.
d — Drizzling.
f — Foggy.
g — Gloomy, stormy looking.
h — Hail.
l — Lightning.
m — Misty.
o — Overcast sky.
p — Passing rain showers.
q — Squally weather.
r — Continuous rain.
s — Snow falling.
t — Thunder.
u — Ugly or threatening weather.
v — Variable weather.
w — Heavy dew.
z — Hazy weather.

Clouds. In the scale for the amount of clouds, 0 represents a clear cloudless sky and 10 a sky entirely overcast. The amount of clouds is recorded in tenths of the sky covered by them. The following are the principal forms of clouds, given in order of their altitude above the earth, beginning with the most elevated.

1. *Cirrus* (Ci.) — Detached delicate, fibrous looking clouds, in the form of feathers, generally white, sometimes arranged in belts converging toward one or two points of the horizon.

2. *Cirro-Stratus* (Ci.-S.) — A thin whitish sheet or a tangled web formation. The sheet formation sometimes causes halos around the sun or moon.

3. *Cirro-Cumulus* (Ci.-Cu.) — *Small* globular masses or white flakes, having no shadows, or very light ones, arranged in groups or lines.

4. *Alto-Cumulus* (A.-Cu.) — *Large* globular whitish or grayish masses, partially shaded, arranged in groups or belts.

5. *Alto-Stratus* (A.-S.) — A thick sheet of grayish or bluish color, showing a brilliant patch in the neighborhood of the sun or moon, but does not produce halos. This form of cloud is similar to the Cirro-Stratus but is only about half as high.

6. *Strato-Cumulus* (S.-Cu.) — Large globular masses or rolls of dark cloud, frequently covering the whole sky, especially in winter. It differs in appearance from the nimbus in this globular or rolled appearance, and does not bring rain.

7. *Nimbus* (N.) — Rain clouds; a thick layer of dark clouds without shape and having ragged edges. Through the opening in these clouds an upper layer of Cirro-Stratus or Alto-Stratus may almost invariably be seen. Loose clouds visible floating at a low level under a Nimbus sheet are *Fracto-Nimbus* (Fr.-N.), called by sailors "*scud.*"

8. *Cumulus* (Cu.) — Wool-pack clouds; thick clouds of which the upper surface is dome-shaped and exhibits protuberances, while the base is horizontal. The true Cumulus has clear superior and inferior limits. It is often broken up by strong winds, and the detached portion is called *Fracto-Cumulus* (Fr.-Cu).

9. *Cumulo-Nimbus* (Cu.-N.) — The thunder-cloud or shower-cloud; heavy masses of clouds in the form of turrets, mountains, or anvils, generally having a fibrous sheet above, and Nimbus beneath. From the base are generally seen showers descending.

10. *Stratus* (S.) — A horizontal sheet of lifted fog; when broken up by wind it is called *Fracto-Stratus* (Fr.-S.).

The State of the Sea is recorded by the following symbols:

B — Broken or irregular sea.
C — Chopping, short or cross
G — Ground swell.
H — Heavy sea.
L — Long rolling sea.
M — Moderate sea or swell.
R — Rough sea.
S — Smooth sea.
T — Tide rips.

(b) THE NAVIGATOR'S NOTE BOOK. The navigator should keep a note book in which to enter all the bearings and observations taken, and all work and calculations connected therewith. It should form a complete history of all navigation performed.

4. REPORTS MADE

Anything of an unusual nature should be reported to the Hydrographic Office at Washington; derelicts sighted, icebergs, any buoy or marker that is suspected to be out of position, any light that is out or that is not showing in accordance with its latest description in the Light List, any unusual meteorological phenomenon, in fact anything that will promote safe navigation, should be reported. Every mariner should have the best interests of the brotherhood at heart.

All navigational information in this country emanates from the Hydrographic Office and it is entitled to the aid of every person who performs navigation.

CHAPTER III

THE VESSEL'S POSITION

THE bearing of an object from a vessel is the direction in which the object is seen from the vessel. It is the angle between the meridian passing through the observer and the object. It is called true, magnetic, or compass, depending upon the meridian of reference chosen. This is explained later.

A LINE OF POSITION is any line on which the vessel's position is known to be, that can be plotted on a chart.

A LINE OF BEARING is any *line of position* obtained from a bearing.

A POSITION POINT is any point on either a line of position or a line of bearing at which the vessel's position is known to be. A position point, generally called a "*fix*," can be obtained by the intersection of two lines of position, two lines of bearings, or one of each.

COMPASS ERROR

VARIATION OF THE COMPASS. Since the earth's magnetic pole in each hemisphere differs in geographical position from the geographical pole, the earth's magnetism will cause the compass needle to point to a spot (the magnetic pole) that is different from the geographical pole (called the *true pole*). Hence the compass needle will point at an angle to the true meridian. The geographical pole lies true North of all points in the Northern hemisphere. The angle that the great circle through the observer's position and the magnetic pole makes with the true meridian (viz: the great circle through the geographical pole) is called the *variation*.

This variation differs for different points on the earth. The variation for any given locality is shown on the charts. A nautical chart always contains the data from which the navigator can find the variation for any locality for any year.

DEVIATION OF THE COMPASS. In addition to the variation, the compass ordinarily has a still further error in its reading. This arises from the effect produced on it by masses of magnetic metal in the vessel itself. Deviation, which is produced by attraction of the compass needle by magnetic iron within the vessel itself, varies for different vessels, different headings of the same vessel, and undergoes change as a vessel proceeds from one geographical locality to another. A table compiled to show the deviation on all headings is called a *deviation table.*

THE COMPASS ERROR is the algebraic sum of the *variation* and *deviation.* As stated before the variation can be obtained from a chart. The deviation is obtained from a deviation table which is constructed by "swinging ship." This and the operation of "compensating the compass" are beyond the scope of the average amateur, but in all ports are mariners who make a specialty of this work. The simplest method of "swinging ship" for deviation, that by ranges, is described on page 49.

From what has gone before it is seen that there are three methods by which bearings and courses may be expressed: (1) *true,* when referred to the geographical, or true, meridian; (2) *magnetic,* when referred to the magnetic meridian; and (3), *compass,* when referred to the meridian in which the compass needle lies.

To convert compass bearings or courses to *magnetic* bearings or courses it is necessary to apply the deviation, corresponding to the vessel's heading, to the compass bearings or courses. Likewise to convert *magnetic bearings* to *true bearings* the variation for the locality must be applied to the magnetic bearings.

The process of applying variation, deviation,

and compass error under all circumstances is one with which the navigator must become thoroughly familiar; these various problems are constantly arising; no bearing can be plotted, or course accurately set, without involving this problem, and careful study of this subject is recommended.

When the effect of a compass error, whether arising from variation or deviation, is to draw the North end of the compass needle to the right the

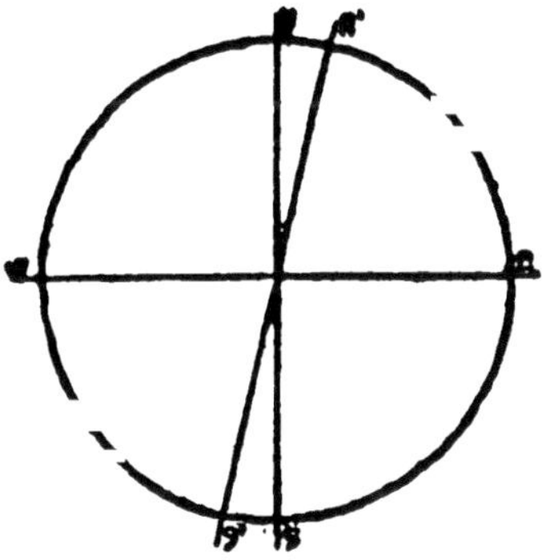

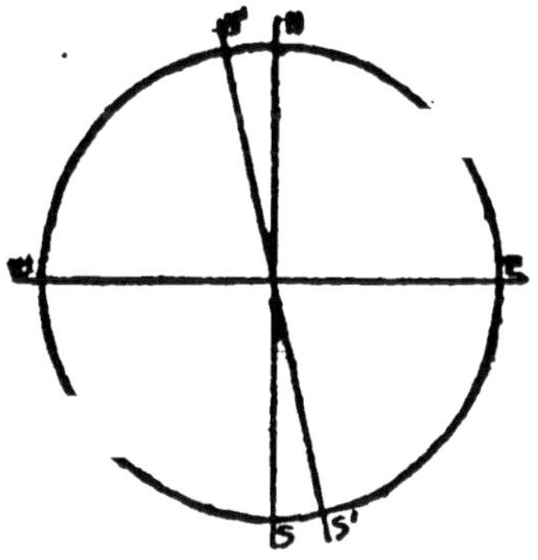

Fig. 7.—Compass Error. Fig. 8.

error is named *East* or is marked +; when its effect is to draw the North end of the needle to the left it is named *West*, or marked —.

Figures 7 and 8 represent, respectively, examples of Easterly and Westerly errors. In both figures consider that the circles represent the observer's horizon, N. and S. being the *true* North and South points in each case. If N.′ and S.′ represent the corresponding points indicated by a compass whose needle is deflected by a compass error, then in Fig. 7, the North end of the needle being drawn to the right the error is Easterly or plus, and in Fig. 8, the North end of the needle being drawn to the left the error is Westerly or minus.

Considering Fig. 7, if we assume the Easterly error to be one point, it is apparent that, if a direction N. by W. is indicated by the *compass*, the *true* direction is N., or one point to the right. Similarly, if the compass direction is N. by E., the true direction is N.N.E., or still one point to the right. If we follow around the whole compass card, the same relation will still hold in every case, the true direction being always one point to the right of the compass direction. Thus, if the compass direction is South, the true direction is S. by W. To understand this consider that you stand in the center of the compass card facing in the direction of the compass reading.

In Fig. 8, assuming the compass error to be one point Westerly, the converse of the above holds true. All *true* directions are one point to the *left* of the corresponding compass directions. Thus, if the compass direction is N. by E., the true direction will be North.

A few simple rules should be remembered when working compass problems:

1. When the *true* bearing is to the *right* of the compass bearing the error is *East.*

2. When the *true* bearing is to the *left* the error is *West.*

3. When applying the compass error imagine yourself standing in the center of the compass card, facing the direction involved in the problem.

4. Deviation plus variation equals compass error. This means the algebraic sum: thus, if variation is 5° E. (+), and deviation is 3° W. (—), compass error is 2° E. (+).

5. To convert from compass to magnetic direction, if the magnetic direction is to the right, the deviation is Easterly. Conversely, to convert a compass direction to a magnetic direction, if the deviation is Easterly, apply it to the right of the compass bearing, if Westerly to the left.

6. To convert from magnetic to true directions, apply the variation to the right of the magnetic direction if the variation is Easterly, and to the left if Westerly.

SWINGING SHIP FOR DEVIATION. Variation for any locality can be obtained from the chart of that locality. To find the compass error on any vessel heading it is necessary to know the deviation on that heading. The simplest way of determining the deviation on the various compass headings of a vessel is to swing ship by the method of ranges.

A Range consists of two well defined and charted objects in line with each other. The direction of the line through these two objects can be ascertained from the chart. Ranges whose magnetic bearings are known have been formed naturally or have been laid out for the aid of navigation in nearly all localities.

To obtain the deviation on various compass headings of a vessel (a deviation table) proceed as follows: Having located a magnetic range, steam across this range on various compass headings (every compass point if time and circumstances permit). Steady on the course for three minutes before crossing the range. Observe the compass bearings (on each heading) of the range when the two objects are in line as observed from the vessel. The deviation for each heading is the difference between the compass bearing of the range on that heading and the known magnetic bearing of the range. RULE.—*The deviation is Easterly when the magnetic bearing is to the right of the compass bearing, and Westerly when the magnetic bearing is to the left of the compass bearing.*

An example will serve to clear up the above: Suppose the magnetic bearing of the range is N.E. (N.45°E.). Steaming across the range the compass bearing on each point of compass heading[1] is taken as follows (columns 1 and 2):

Column 1 Compass Heading	*Column 2* Compass Bearing of Range	Bearing of Range *Column 3* Magnetic	*Column 4* Deviation
N	N 47° E	N 45° E	2 W
N by E	N 48° E	"	3 W
NNE	N 49° E	"	4 W
NE by N	N 47° E	"	2 W
NE	N 45° E	"	0
NE by E	N 43° E	"	2 E
ENE	N 42° E	"	3 E
E by N	N 41° E	"	4 E

Record in column 3 the magnetic bearing of the range. To obtain column 4 take the numerical differences between columns 2 and 3 and mark the results East or West according to the rule given above.

Before Swinging Ship see that all portable metal objects, especially those near the compass, are secured in the positions habitually occupied by them at sea. A change in the position of metal objects near the compass will affect its deviation. See that the vessel is on an even keel. This rule also applies when taking bearings. Steam across the course slowly, having enough headway to keep a steady course.

[1] Only eight compass points are taken for demonstration.

The navigator should swing ship when practical before any long trip, or after laying in port any very long interval. It is not necessary to obtain the deviation on every compass point. If a table of deviations on every other point is constructed the deviation on intermediate points can be interpolated.

THE VESSEL'S POSITION

FINDING THE VESSEL'S POSITION, when piloting, resolves itself into four general cases:

1. To get the vessel's position when one object of known position is in sight.
2. To get the vessel's position when two objects of known position are in sight.
3. To get the vessel's position when more than two objects of known position are in sight.
4. To get the vessel's position when no objects of known position are in sight.

Case 4 is a method of soundings and as such is discussed in Chapter 5 under that heading.

(1) TO GET THE VESSEL'S POSITION WHEN ONE OBJECT OF KNOWN POSITION IS IN SIGHT.

(a) *By bearing and distance.* Take a compass bearing of the object. Convert this bearing to magnetic or true bearing by applying deviation or compass error. Plot this line of bearing on the chart by means of parallel rulers, using the magnetic or true compass rose, depending upon whether the bearing has been converted to a magnetic or true bearing. The vessel lies somewhere on this line. If the distance from the object can be obtained, an arc of this length with the object as a center will intersect the line of bearing at the vessel's position.

The distance may be estimated, but this is of doubtful value. An estimated distance may be verified by a sounding, if the bottom is very irregular so that a characteristic sounding may be obtained.

The distance from an object may be found by the vertical angle method if the height of the object is known, and the angle the object subtends at the observer can be measured. The heights of all lighthouses are given in the Light List. Measure the angle subtended by the lighthouse by means of a sextant. Now

$$d = .567 \frac{h}{O}$$

where d = distance required in knots
h = height of lighthouse
O = angle measured, in minutes of arc.

Example: — A lighthouse 150 feet high subtends an angle of 12′; find the distance.

$$d = \frac{150}{12} \times .567 = 7.09 \text{ miles (nautical)}$$

Table 33 of Bowditch's Tables (Edition of 1913) can be used to solve problems of the above kind. Enter the column corresponding to the height of the object. Find the angle nearest to that measured, and on the same line (in the first column) is found the distance of the object.

(b) *By Two Bearings of a Single Object and the Course and Distance Run in the Interval.* In

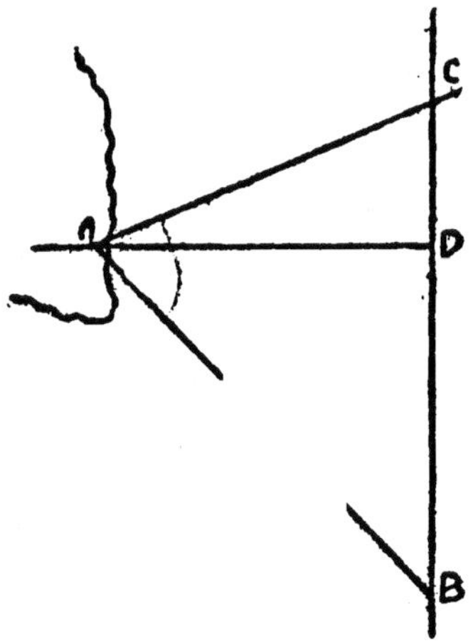

FIG. 9.—Position by two bearings and interval run.

Figure 9, assume A is the object on which bearings are taken. The vessel being at B, the bearing BA is taken and the patent log is read. The vessel then steers a course BC and at C the bearing CA is taken and the patent log is read. The difference of patent log readings is the distance BC. (Note: If no patent log is carried the distance BC can be computed by knowing the speed at which the vessel is sailing and the time it takes to run the distance BC.) From the course steered and the bearings BA and CA the angles B and C can be computed and BAC = 180 — (B + C). The problem is one of plane trigonometry, given two angles and one side of a plane triangle. It can be solved in three ways:

(A) by plane trigonometry,
(B) by Bowditch's Tables,
(C) by graphic chart work.

(A) This method is little used, the last two being far simpler. By plane trigonometry,

$$AB = \frac{\sin C}{\sin A} \times BC$$

$$AC = \frac{\sin B}{\sin A} \times BC$$

Dropping the perpendicular AD from A to BC, the distance of the object when abeam is

$$AD = AC \sin C.$$

(B) Tables 5A and 5B, Bowditch, are compiled for an inspection solution of this problem. The arguments in these tables are the difference between the course and the first bearing, and the difference between the course and the second bearing. The differences in Table 5A are in quarter points, and in Table 5B are in degrees. The factors in the columns are for a distance run of one mile. So far as the factors are concerned it is immaterial whether the courses and bearings are compass, magnetic, or true, so long as they all refer to the same meridian.

To use the tables: Enter the column corresponding to the difference between the course and the first bearing. Run down this column until the line is reached corresponding to the difference between the course and the second bearing. On this line are found two factors. Multiply the run between bearings by the first factor and the result is the distance of the object at the time of the second bearing. Multiply the run by the factor in the second column and the result is the distance when the object is abeam.

Example: A vessel heading 250° (p.s.c., per standard compass), had a lighthouse bearing 202° (p.s.c.); after a run of 10 miles the same light bore 130° (p.s.c.). Find the distance at the time of second bearing, also when abeam.

Difference between course and first bearing .... 48°.

Difference between course and second bearing ..120°.

Table 5B, factor first column = 0.78 and distance at second bearing = 10 × 0.78 = 7.8 miles.

Table 5B, factor second column = 0.68 and distance when object is abeam = 10 × 0.68 = 6.8 miles.

(C) There is a graphic method of solving this problem that is preferred by some navigators to the factor method. Draw upon the chart the lines CA and BA, Figure 10, passing through the object and having the direction of the two observed bearings; set the dividers to the distance run, BC; lay down the parallel rulers in a direction parallel to the course steered and move them toward or away from the observed object until such a point

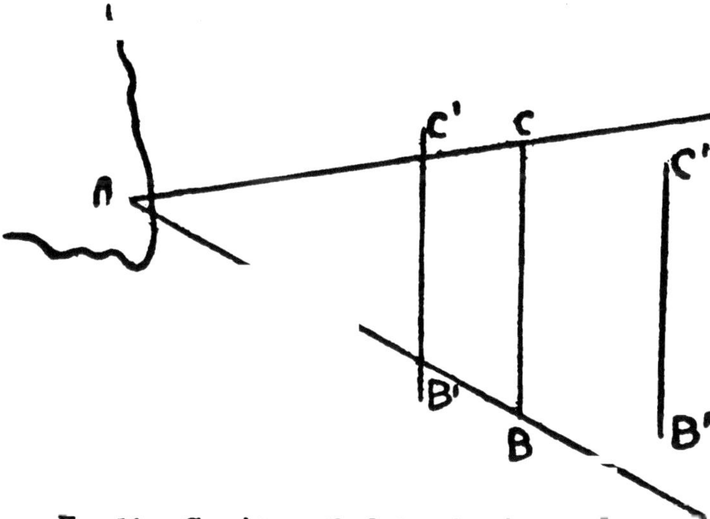

FIG. 10.—Graphic method, two bearings and run.

is found that the distance between the lines of bearings is equal to the distance between the points of the dividers. In the figure this occurs when the rulers lie along the line BC, and therefore B represents the vessel's position at the time of first bearing, and C its position at the time of second bearing. For any other positions B″C″, B′C′, the condition is not fulfilled.

(D) *Special Problems.* There are many special applications of the above method of obtaining the vessel's position, that are so simple as to become mere mental problems. Some of these are given here:

Bow and Beam Bearings. If the first bearing is is taken broad off the bow (45° from ahead) and

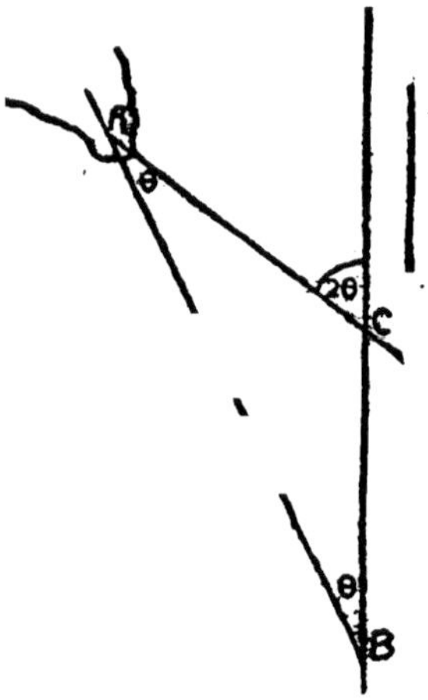

FIG. 11.—Doubling Angle on the Bow.

NOTE: In Figure 11, suppose the angle observed between the course and the bearing at B is θ, and at C is 2θ. Then the angle BAC equals θ and AC equals BC.

the second bearing is taken on the beam (90° from ahead), then it is apparent from inspection that the distance at the time of the second bearing equals the distance run. Likewise, if the first bearing is taken abeam (90° from ahead) and the second is taken on the quarter (135° from ahead), then the distance when abeam is equal to the distance run. The distance from the object when on the bow and quarter is 1.4 times the distance run.

Doubling the Angle on the Bow. Take a bearing of an object when forward of the bow; read the patent log. Take a second bearing of the object when its angular distance from ahead is twice as great as it was at the time of the first bearing; read the patent log. The distance of the object from the vessel at the time of the second bearing is equal to the distance run between bearings.

The 26½° Rule. If the first bearing is taken when the object is 26½° from ahead, and the second bearing is taken when the object is 45° from ahead, then the *distance at which the object will be passed abeam* is equal to the run between the two bearings.

The Seven-tenths Rule. If bearings are taken of an object when two and four points on the bow, seven-tenths (0.7) of the run between bearings is the distance at which the object will be passed abeam.

The Seven-thirds Rule. If bearings are taken of an object when two points forward of the beam (67½° from ahead), and when abeam, seven-thirds (7/3) of the run between bearings is approximately the distance of the object when abeam.

If bearings are taken of an object at 22½° and

26½° from ahead, then ⅞ of the distance run between bearings is approximately the distance at which the object will be passed abeam.

The method of obtaining the position by two bearings on the same object is very useful, for frequently it is necessary to locate one's position when only one landmark is in sight. A good navigator never misses an opportunity to check his position by a bow and beam bearing on every well charted object passed.

It must always be borne in mind that the results from this method will be incorrect unless the "course and distance made good over the ground" are correctly estimated. Bad steering, cross currents, or leeway of the vessel will cause inaccuracy in the estimated course, and a current with or against the ship, or inaccuracy in the distance logged will affect the estimated distance. A current with the vessel will give a position closer to the charted object than the actual position, and a current against the vessel will give a position farther away than the actual distance. If the local current is known, allowance can be made for it when working this problem.

(2) To Get the Vessel's Position When Two Objects of Known Position Are in Sight. (*Cross Bearings of Two Charted Objects.*)

Select two well charted objects whose respective

bearings differ as nearly as possible by 90°. Take the compass bearings of these objects when the vessel is on an even keel, and as quickly as possible one after the other. Correct the compass bearings so that they will be either true or magnetic, according to the compass rose to be used in plotting them, applying compass error to ob-

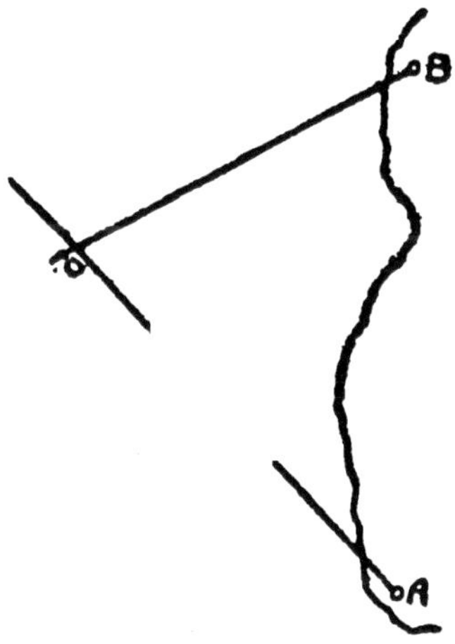

Fig. 12.— Position by cross bearings.

tain true bearings, and deviation only to obtain magnetic bearings.

By means of parallel rulers draw through the objects lines in the respective directions that each was observed to bear. As the vessel's position is known to be on each of these, it follows that it must be at the intersection. In Figure 12, if A and B are the objects and OA and OB are the lines passing through them in the observed directions, then the vessel's position is at O, their intersection. The solution of this problem is greatly facilitated by the use of Koch's Plotter, which is described with its uses at the end of this chapter.

If possible, objects selected for *cross bearings* should subtend an angle at the vessel of at least 30° and less than 150°. As the angles become smaller and greater than these limits, small errors of observation give larger errors in the results. Near objects are preferred to distant ones. When finding the "fix" by cross bearings, take the bearing of the object nearest ahead or astern first, as that object will change its bearing less during the interval between bearings than one more nearly abeam.

(3) To Get the Vessel's Position When More than Two Objects Are in Sight.

(a) *Cross Bearings.* The method of cross bearings on two objects has been explained above. If a third well charted object is in sight, its bearing should be taken and plotted in the same manner to be a check on the other two. If this line of bearing intersects at the same point, O in Figure 12, it verifies the accuracy of the "fix." If not it indicates an error, either in the observations, plotting, or the application of the compass error. Should the three bearings form only a very small triangle at their intersection, the center of this triangle may be taken as the vessel's position.

(b) *Three Point Problem.*[1] This is the most accurate method of obtaining a fix. Three objects on the chart are selected and the angles between them are measured by a sextant. In lieu of a sextant the bearings of three objects may be taken and the angles between them may be computed therefrom. The position is plotted by a *Three Arm Protractor* (a description of this instrument is omitted because the Koch Plotter described later will perform all of its functions). To plot, given the angles, set the right and left angles on the instrument and then move it over the chart until the three bevel edges of the arms pass respectively and simultaneously through the three objects. The center of the instrument will then mark the fix, which may be marked by a pencil point through the center hole of the protractor.

(4) DANGER ANGLES. When coasting the navigator will often desire to pass sunken rocks or dangerous shoals or sunken obstructions at a minimum distance. This is done by use of the *Danger Angle*, of which there are two kinds, horizontal and vertical. The former, which is prefer-

[1] This method is used extensively in marine surveying. It requires the use of a sextant and is of little value to motor boats.

able, is used when two charted objects making a fair angle are available, and the latter when only one such object is available.

The Horizontal Danger Angle.[1] In Figure 13, suppose AMB a portion of the coast along which the vessel is coasting, on the course CD; A and B are two well charted objects; S and S′ are two outlying shoals or reefs that must be avoided. To pass outside the danger S′ proceed as follows: (the construction is shown in full lines) take the middle point of the danger S′ as a center, and with a radius equal to the distance from the danger that it is desired to pass, describe a circle (the small full circle). Now pass a circle (the large full circle) through A and B tangent to the small circle. Measure the angle AEB, set the sextant to this angle. Now since AB subtends the same angle at all points of the circle AEB it is apparent that as long as AB does not subtend an angle greater than AEB, the vessel will be outside the circle AEB, hence clear of the danger S′

To avoid the danger S, passing inside it, proceed in a similar manner as follows (construction in broken lines): Take the middle point of the danger S as a center, and with a radius equal to

1 This requires the use of a sextant. The knowledge is unnecessary but is inserted for the use of the navigator who possesses a sextant.

the distance it is desired to pass the danger, describe a circle (the small broken circle). Through A and B draw a circle tangent to the first circle

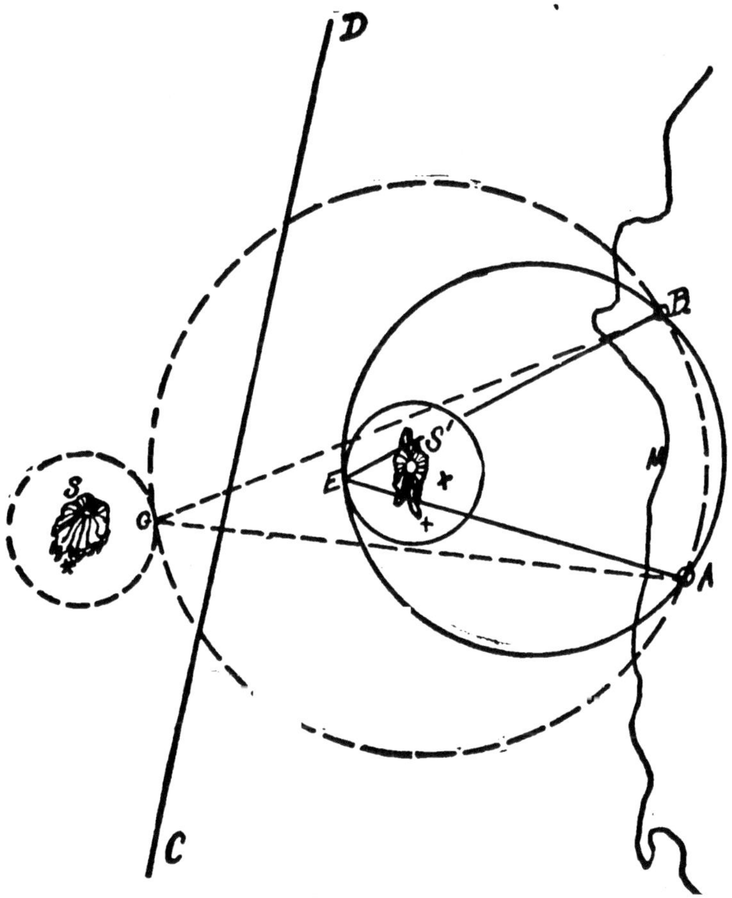

FIG. 13.— Horizontal Danger Angle.

(large broken circle). Measure the angle AGB with a protractor and set the sextant to this angle. Now as long as AB subtends an angle greater than AGB the vessel will pass inside the danger S. By combining the two methods, the ship will pass safely between the dangers as long as the angle subtended by AB is less than AEB and greater than AGB.

Vertical Danger Angle, Figure 14. This

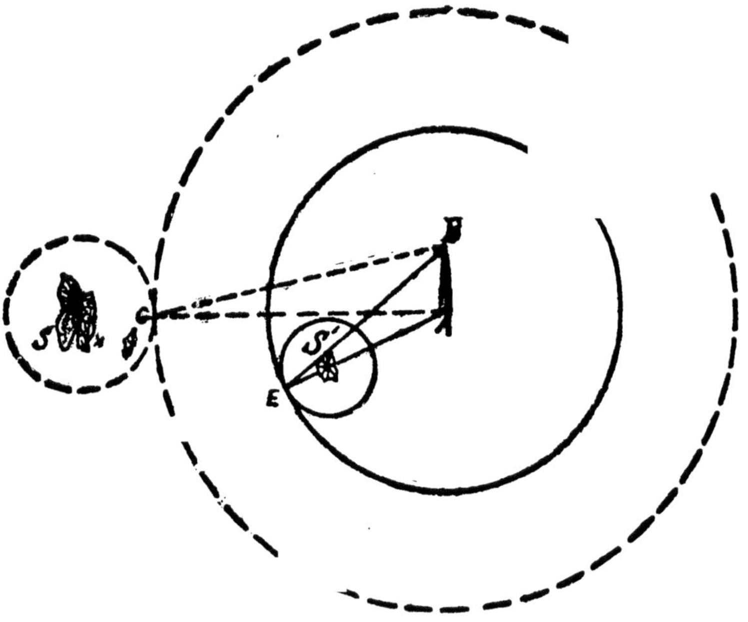

FIG. 14.—Vertical Danger Angle.

method is based on similar principles to the horizontal danger angle, the construction being a little different. An object of known height and distance, such as a lighthouse, is used. Lighthouses are measured from mean high water to center of lantern. Allow for stage of tide if there is much range. Draw the safe circle with S′ as a center. With the object as a center draw a second circle tangent to the outside of the first circle at E. Measure the distance AE on the chart. With the distance AE and the known height AB the angle AEB can be computed. Now as long as the angle subtended by AB is smaller than the angle AEB the vessel will pass safely outside of S′. By similar argument, as long as the angle subtended by AB is greater than AGB the vessel will pass safely inside the danger S.

Danger Bearings are useful when coasting to warn a navigator by *one compass bearing* when the course is leading into danger. Suppose a vessel is steering a course, as shown in Figure 15, along a dangerous coast with an outlying reef as shown, with only the landmark A as a guide. Draw through A a line AX that will clear the danger all along. Note its direction by the compass rose. Take frequent bearings of A. As long as the bearings YA and ZA are to the right of XA the vessel is on the safe side of XA and clear of danger. If a bearing taken is to the left of XA the vessel must steer off shore.

Although the object in sight may be so nearly ahead as to be valueless for a definite fix, this method may be employed to keep the vessel out of danger.

Ranges. A range consists of two well charted objects in line that can be used as a navigational

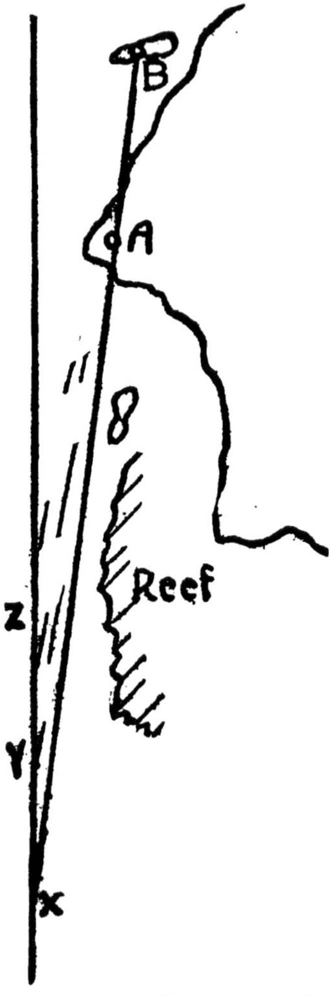

Fig. 15.—Danger Bearing and Range.

aid. Thus in the above problem, if another well charted object lies in the prolongation of AX as at B, Figure 15, the line XAB is called a range. It is unnecessary to take the bearings YA and ZA in this case, because as long as the rear object is seen to the left of A (called an "open range"), the vessel is safe. Ranges are used a great deal in river and harbor work.

The course to be steered is marked by two objects, which when kept in line indicate a safe course to steer. Sometimes a danger will lie on one side of the range with good water on the other. This constitutes in effect a danger bearing. When taking cross bearings for anchoring it may happen that one of the bearings will be on a range, which can be plotted directly on the chart without observing its direction. Anchoring on ranges in open harbors is a very common practice.

If in a strange locality, the navigator should observe and compare the compass bearings of all ranges with the bearings indicated on the chart in order to make certain of their identity.

ROUNDING AN OBJECT AT A GIVEN DISTANCE. To steer an arc course around a light and keep it at a given distance without the use of "fixes," *provided there is no current,* stand on the first course until the light is at the desired distance. Immediately bring the light abeam and steer this new course until the light is one-half point abaft the beam. Now change course until the light bears one-half point forward of the beam and steer this new course until the light bears one-half point abaft the beam again and repeat. By this method the vessel travels on a polygon, the inscribed circle of which has a radius of the desired distance. The number of sides of the polygon may be indefinitely increased, so that the light may be rounded by frequently changing the course just enough to keep the light abeam.

THE KOCH PLOTTER. Reference has been frequently made to this instrument and it is described here instead of in Chapter 2 because its use is peculiarly applicable to the problems here presented. It can perform all the functions of the compass rose, parallel rulers, three arm protractor, etc. By using this instrument all the above can be done away with. The writer has frequently used it and would not go to sea without one. It facilitates the solution of all piloting problems that involve bearings, angles, or courses. It is simple in construction as the following description indicates.

All materials except bolts and washers are transparent. There are two celluloid plates, one 7″ square plate with two series of lines perpendicular to each other, the lines of each series being about ½″ apart, the other a 7½″ circular disc, marked on its circumference in degrees. These are centered on a hollow bolt of brass and can be clamped together with any degree of friction desired. Three arms are placed so as to revolve around the hollow bolt and can be clamped together in any position desired. The following are

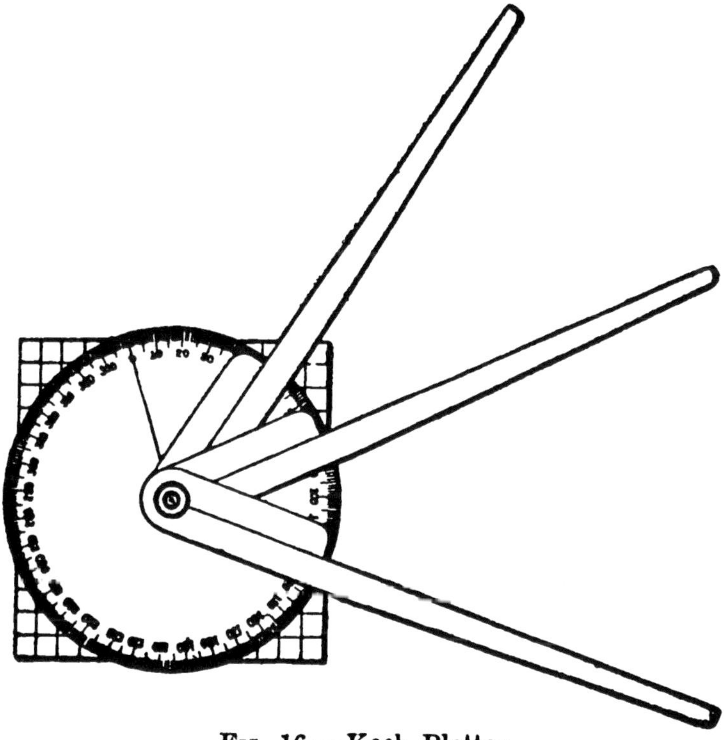

FIG. 16.—Koch Plotter.

a few of the problems that can be solved by this ingenious instrument. (It should be noted that no parallel ruler, or other instrument, is used, and reference is not made to the compass rose on the chart.)

(1) *To set a given course from a given position.* Revolve the zero mark of the disc to the East or West of the True North and South line of the square an amount equal to the compass error. Set one arm to the desired compass course. Lay the plotter with its center at the given position. Revolve the plotter about this position as a center until a vertical line on the square coincides with a meridian, or a horizontal line with a parallel of latitude. The course now lays along the bevel edge of the set arm and can be drawn if desired.

(2) *To find the course between two positions.* Set the disc for compass error as before. Center the disc on one position and revolve it as before. When a vertical or horizontal line of the square is in coincidence as before, swing one arm so its bevel edge passes through the second object. Read off the course where this bevel edge cuts the degree marked disc.

(3) *To plot the position from two bearings.* Set the disc for compass error as in (1). Set the two arms to the two compass bearings on the disc. Clamp the arms. Manipulate the plotter on the chart so that the bevel edges of these two arms pass through the two observed objects and the vertical lines on the square are parallel to the meridians. With a pencil the desired position

can be marked through the hollow central bolt. The third arm can now be swung to mark a new course or to get a desired compass course to any object.

(4) *To plot the position by three compass bearings or the angles between three objects.* In this case the lines on the square plate are *not* brought into parallelism with the meridians on the chart because the compass error is disregarded. Set the disc to zero on the reference line of the square card. Set the three arms to the compass bearings, by the degrees on the disc. When plotting by observed angles, set one arm at zero, the second arm by degrees on the disc at the left angle reading, and the third arm by degrees on the disc to a reading equal to the sum of both angles.

Having set the three arms by either of the above methods, manipulate the plotter on the chart until the bevel edges of the arms intersect the three objects on the chart, the center arm being on the center observed object. Mark the position by inserting a pencil in the hollow central bolt.

(5) *To get the compass error from the compass bearings on three objects.* Proceed as in (4). Set the arms to the compass bearings and get the position. With the disc set at zero on the reference line of the square the lines on the square

will not now be in parallelism with the meridians on the chart unless the compass error is zero. If not parallel, the compass error can be found as follows: Keep the disc in its position on the chart and revolve the square until its lines are parallel to the meridians of the chart. The angle that now shows between the reference line of the square and the zero of the disc is the compass error.

Many other problems can be solved as they arise. The principle of this plotter is the portable compass rose. When the square card has its lines parallel to a meridian or parallel of latitude, the disc is in effect a compass rose. By adjusting the disc to zero, deviation, or compass error, the compass rose indicates directions per compass, magnetic, or true, respectively. The instrument is a combination of compass rose, parallel rulers, and three arm protractor.

CHAPTER IV

DEAD RECKONING

CORRECTING the Compass Course. When navigating by dead reckoning all courses are reduced to true courses. The method of correcting the compass course to obtain the true course is given in Chapter III.

The Compass error is the *algebraic* sum of the *deviation* and *variation.* If the compass error is *East,* apply it to the *right* of the *compass course* to get the true course; if *West,* apply it to the *left* of the *compass course* to get the true course. When applying this correction imagine that you stand in the center of the compass card. Some compasses are marked from 0° to 360°, while others are marked from 0° to 90° from the North and South to the East and West. Eight problems follow to illustrate compass correction. The first solution in each case is for a compass of the first type, the second solution (in parenthesis) for a compass marked in the second manner.

Prob. 1. Compass Course 75° (N75°E), Compass error + 5° (5° E).

Find true course. Answer 80°. (N80°E.)

Prob. 2. Compass Course 75° (N75°E), Compass error — 5° (5°W).
Find true course. Answer 70°. (N70°E.)
Prob. 3. Compass Course 150° (S30°E), Compass error + 10° (10°E).
Find true course. Answer 160°. (S20°E.)
Prob. 4. Compass Course 150° (S30°E), Compass error — 10° (10°W).
Find true course. Answer 140°. (S40°E.)
Prob. 5. Compass Course 220° (S40°W), Compass error + 4° (4°E).
Find true course. Answer 224°. (S44°W.)
Prob. 6. Compass Course 220° (S40°W), Compass error — 4° (4°W).
Find true course. Answer 216°. (S36°W.)
Prob. 7. Compass Course 305° (N55°W), Compass error + 6° (6°E).
Find true course. Answer 311°. (N49°W.)
Prob. 8. Compass Course 305° (N55°W), Compass error — 6° (6°W).
Find true course. Answer 299°. (N61°W.)

The Sailings. When a vessel sails from one place to another on the earth's surface, the computations connected therewith involve five quantities, viz.: the *Course*, the *Distance*, the *Difference of Latitude*, the *Departure*, and the *Difference of Longitude*. Solution of problems involving these quantities is called *Sailings*. There are many kinds of *Sailings* employed by navigators, of which the following is a list.

1. Plane Sailing.
2. Traverse Sailing.
3. Spherical Sailing.

4. Parallel Sailing.
5. Middle Latitude Sailing.
6. Mercator Sailing.
7. Great Circle Sailing.
8. Composite Sailing.

The last three will not be considered as they are used in cases where the distance sailed is very large, and do not come within the scope of this work. *Middle Latitude* sailing is the one most used in *Dead Reckoning.* An understanding of the first four, however, is necessary to an understanding of Middle Latitude sailing, and they are here discussed to lead up to Dead Reckoning. Although the problems of Dead Reckoning can be solved by trigonometry, in practice they are solved by the use of Tables 1 and 2 of Bowditch's Useful Tables.

The curved line joining any two places on the earth's surface and cutting all meridians at the same angle is called the *Rhumb Line.* The constant angle which this line makes with the meridians is called the *Course;* the length of the line between any two places is called the *Distance.*

PLANE SAILING. For the moment, suppose the curvature of the earth is neglected. In Figure 17, T is the point of departure, T′ the point of destination, TT′ is the rhumb line. Draw Tn and T′n, forming a right angle. Tn is the meridian and T′n is the parallel of latitude. nTT′ is the course, being the angle the rhumb line makes with the meridian. TT′ is the distance (Dist.), being

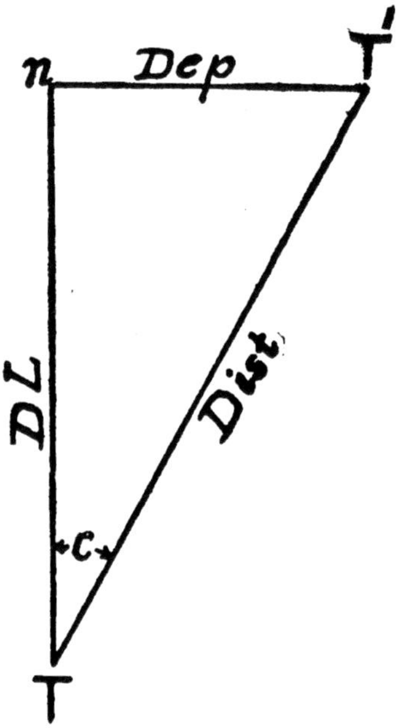

FIG. 17.— Plane Sailing.

the length of the rhumb line. Tn is the difference of latitude (DL) and T′n is the departure (Dep). Then from the right triangle, TT′n, we have the following formulæ:

$$\text{Sin C} = \frac{\text{Dep}}{\text{Dist}}$$

$$\text{Cos C} = \frac{\text{DL}}{\text{Dist}}$$

$$\text{Tan C} = \frac{\text{Dep}}{\text{DL}}$$

By use of these equations all problems that arise in plane sailing can be solved, trigonometrically.

A much simpler method of working the sailings is by the use of Tables 1 and 2, Bowditch. These are commonly called the Traverse Tables. The tables are simply solutions of right triangles. Referring to any page of Tables 1 or 2, at the top (or bottom) of the page is the course or angle, C. The three quantities found on any one line are the three sides of the triangle, Distance, Latitude, and Departure. By entering this table with any two known quantities, the other two can be found by inspection. A few examples will make this clear.

Example 1. A ship sails SSW, 250 miles. Required the difference of latitude and departure.

Enter table 1, at course SSW. It occurs on top of page so take names of columns from top. Under the column marked Dist. look for 250. On the same line is the solution under the columns marked Lat. and Dep.

Diff. of Lat. = 231 miles = 231′ = 3°—51′ S.

Dep. 95.7 miles W.

Example 2. A vessel sails S80 E, 190 miles. Find the difference of latitude and departure.

All courses are figured from North as zero right around

the compass in a clockwise direction. Therefore S80 E is the equivalent of N100 E, or course 100. Enter Table 2 at the course 100. This occurs at the *bottom* of the page so take all column names from the *bottom*. Opposite Dist. 190 we find

Dep = 187.1, Lat = 33, therefore

DL = 33′S.

Dep = 187.1 miles E.

Example 3. A vessel has sailed 115 miles to the North and 155 miles to the West. Required the Course and Distance sailed (using table 1).

In this case Lat = 115N and Dep = 155W. Enter table 1 and find a course where 115 and 155 are found abreast each other in the columns marked Lat and Dep respectively. This occurs most nearly at 4¾ points; the angle is taken at the bottom because the proper names of the columns occur there. Since the Lat is N and the Dep is W the

Course = NW¾W. The distance is found on the same line as 115 and 155, and in this case is 193 miles.

Example 4. A vessel sails 67 miles to the North and 52 miles to the West. Required the Course and Distance, using table 2.

In this case Lat = 67N and Dep = 52W.

Enter table 2 and find the page where the known quantities occur on the same line. This occurs most nearly under 32, and the distance is 79. Since the proper column names occur at the top of the page, the result is

Course = N32°W = 328°.

Distance = 79 miles.

Traverse Sailing. So far we have considered the case where only one course and one distance is involved. If the vessel sails several different courses and distances, these can all be reduced to one *equivalent* course and distance by the method of *traverse sailing*. This is done by finding the difference of latitude to the North or South and the departure to the East or West for each course and distance and then taking the algebraic sum of these differences of latitude and departure to find the equivalent course and distance.

Suppose in Fig. 18 that the vessel has sailed the distance AB on the course OAB and then sailed

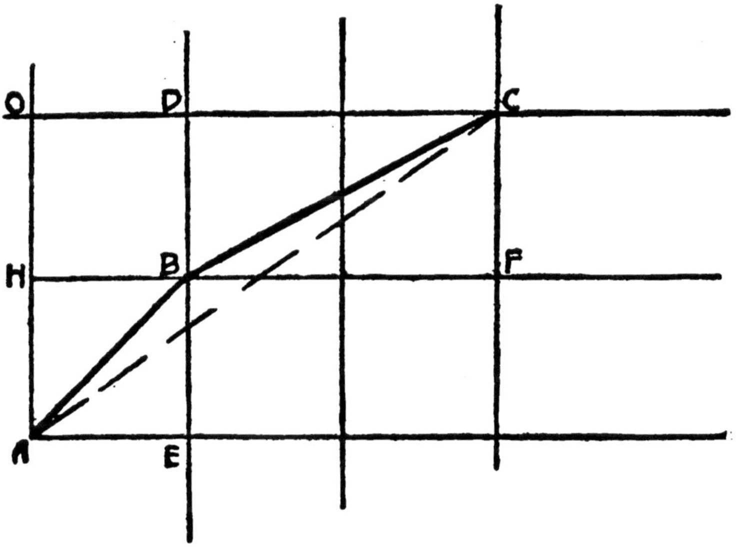

FIG. 18.—Traverse Sailing.

the distance BC on the course DBC. Its final position, by diagram, is the same as though it had sailed the distance AC on the course OAC.

Considering the first course and distance, HA equals the corresponding difference of latitude and HB the corresponding departure; considering the second course and distance DB equals the corresponding difference of latitude and DC the corresponding departure. Add the differences of latitude.

$$HA + DB = AO.$$

and add the departures

$$HB + DC = OC.$$

but AO and OC are the difference of latitude and departure respectively for the course OAC and distance AC.

To sum up: Starting with an initial position, A, and given several courses and corresponding distances, the final position C is arrived at by adding the several differences of latitude and several departures and with these two sums proceeding as in plane sailing.

A tabular form is used for the work as follows:

Example. A vessel sails SSW, 12 miles; SW, 40 miles; ESE, 5 miles; E by N, 60 miles; SE by S, 12 miles. Find the course and distance made good.

This is solved in a tabular form as follows: In column 1 place the courses, in column 2 the distances, in columns 3 and 4 the Lats, according as they are North or South, and in columns 5 and 6 the Deps, according as they are East or West. Next take the difference between the sums of columns 3 and 4, and the difference between the sums of columns 5 and 6, in each case retaining the names of the larger quantity. With these differences as the *equivalent* Lats and Deps enter table 1 and get the *equivalent* Course and Distance.

Parallel Sailing. Thus far the spherical form of the earth has been ignored. It has only

Course	Dist	Lat		Dept	
		N	S	E	W
SSW	12		11.1		4.6
SW	40		28.3		28.3
ESE	5		1.9	4.6	
E by N	60	11.7		58.8	
SE by S	12		10.0	6.7	
		11.7	51.3	70.1	32.9
			11.7	32.9	
SE 1/4S	53.5		39.6	37.2	

FORM FOR TRAVERSE SAILING.

been considered as a plane surface, and *Difference of Longitude* has not been taken into account. Problems involving Differences in Longitude (abbreviated DLo) are solved by *Spherical Sailing*, of which *Parallel Sailing* is the simplest form. When a vessel sails upon an East or West course a certain distance, this distance is the departure. Converting this departure into degrees and minutes of longitude, or vice versa, is done by parallel sailing.

Suppose in Figure 19, T and T′ are two places in the same latitude; P, the adjacent pole; TT′ the arc of the parallel of latitude between the two places (is the departure); MM′, the corresponding arc of the equator intercepted between the meridians passing through the two places (is the difference of longitude); CMM′ is the plane of the equator; CP is the earth's axis; TOT′ is a plane

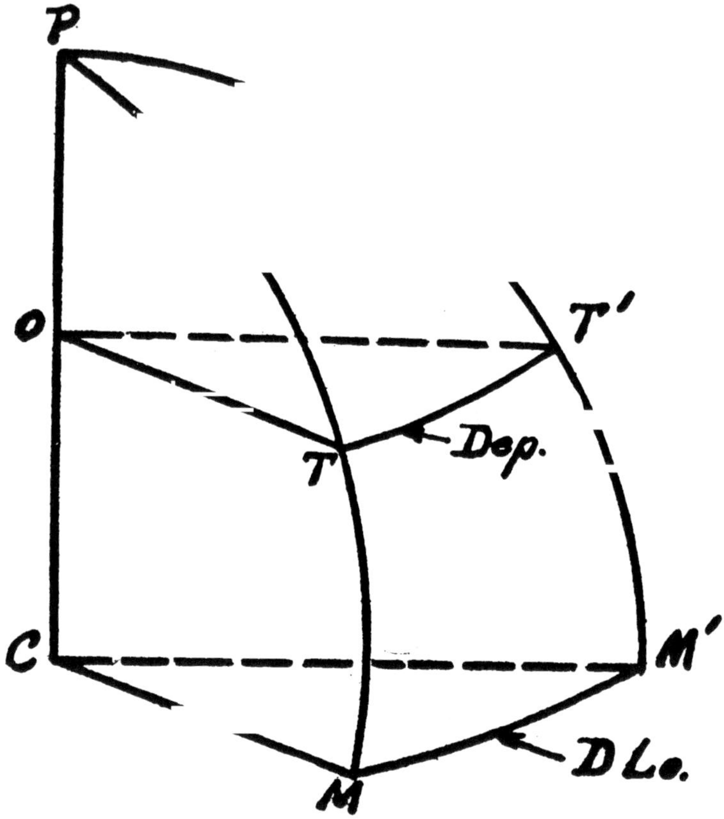

FIG. 19.— Parallel Sailing.

through the parallel of latitude TT′. Now by construction

TT′ = Dep   MM′ = DLo

It can be shown trigonometrically that

DLo = Dep sec L

This formula expresses the relation between departure in miles and difference of longitude in minutes. Problems in interconversion of Dep and DLo may be worked by the above formula. A much simpler method is the use of Traverse Tables. Since the Traverse Tables are based on the formula DL = Dist cos C, we may substitute the departure for the column marked Lat, and the difference of longitude for that marked Dist, and the Latitude for the courses marked at the top and bottom of the page. The tables can then be used to these conversions.

Example. A vessel in Latitude 40° sails due East 167 miles. Required the difference of longitude.

Enter table 2 at course marked 40°. Under column marked Lat find 167. Pick out the number opposite 167 in the column marked Dist. This is the required difference of longitude in minutes, in this case, 218′ or 3° — 38′E.

Example. A vessel in Latitude 20°—30′ sails due West a distance of 140 miles. Required the difference of longitude. In this case, since these tables are made only for even degrees, the result must be picked out for Lats. (courses in Table) 20° and 21°. In this case the differences of longitude are

for 20°......149′ }
for 21°......150′ } The mean is 149.5′
DLo = 149.5′ = 2° 29.5′W

Middle Latitude Sailing. When a vessel sails on other than an East or West course on a parallel of latitude, its latitude is constantly changing and the *Parallel Sailing* method of interconverting departure and difference of longitude must be modified. In Figure 20, T is the point of departure; T′ the point of destination;

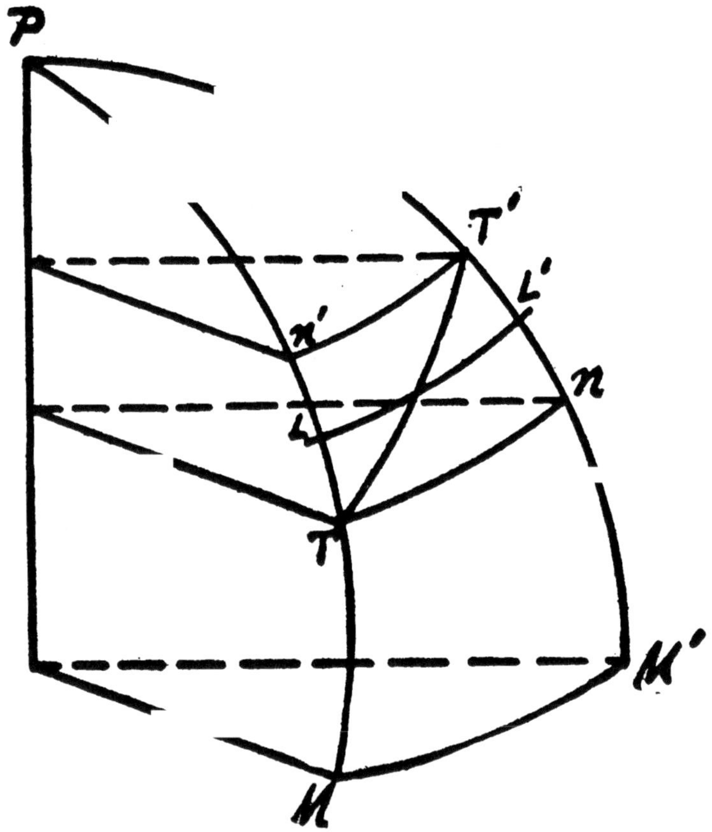

Fig. 20.—Middle Latitude Sailing.

P, the earth's pole; TT′ the rhumb track; n′TT′, the course; Tn and n′T′, the respective parallels of latitude: MM′, the equator. By construction the difference of longitude is MM′. Now draw a parallel of latitude LL′ halfway between Tn and n′T′. The change in latitude between T and T′ is Tn′. Call the latitude of T "Lat Left," and the latitude of T′ "Lat arrived at," and the latitude of LL′, "Middle Lat."

Had the vessel made all of its change of longitude along the line Tn, then DLo = Dep sec (Lat Left). If it had all been made along the line n′T′, then DLo = Dep sec (Lat arrived at). Since the change was made along intermediate lines, neither formula is applicable. It can be shown mathematically that for an ordinary day's run of a vessel the formula DLo = Dep sec (Middle Lat) is correct. The method of conversion based upon this formula is called Middle Latitude Sailing.

Having found the latitude arrived at, from the latitude left and the difference in latitude, the mean of the two is taken and the solution of the problem becomes the same as the solution for *Parallel Sailing*, substituting the *Middle Latitude* for the single Latitude used therein.

Example. A vessel in Latitude 42° 30′N, Long. 58° 51′W, sails SE by S, 300 miles. Required the latitude and longitude arrived at.

Entering Table 1, Bowditch, Course SE by S, Dist 300, we find Lat 249.4 S (4° 09.4′), Dep 166.7E. Proceeding

Lat left	...42° 30.0′N	Long left	............58° 51′ W
DL	.... 4° 09.4′S	Diff. Long (DLo)	.219.2′ 3° 39.2′E
Lat arr at	38° 20.6′N	Long. arr. at	........55° 11.8′W
	42° 30.0′		
	2)80° 50.6′		
Mid. Lat	..40° 25.3′		

First find the new latitude, then take the mean of the first and second latitudes for the *Middle Latitude* (40° 25.3′). Enter Table 2 with the middle latitude as a course.

With Mid. Lat. 40° as course, DLo (under Dist. column) corresponding to a Dept (under Lat column) of 166.7 is 217.6; with Mid. Lat 41° as course, DLo is 220.9. The mean value, corresponding to 40½° is 219.2.

DEAD RECKONING is the process of determining at any instant the vessel's position by applying the course and distance run from any previous well determined position. Having once determined the vessel's position, the position at any subsequent time may be found by applying the difference in latitude and difference in longitude obtained by the method of *traverse sailing* described in this chapter. Positions thus obtained are called Dead Reckoning (abbreviated D.R.) positions, or positions *by account.*

Positions by dead reckoning are not as accurate as those by observation because they may be influenced by incorrect estimate of distance run, bad steering, incorrect compass error, unknown currents, etc., and for this reason every opportunity that presents to fix the position by bearings should be grasped.

To obtain the vessel's position by dead reckoning it is necessary to have some previous well determined position. When a vessel leaves port, its position is always accurately determined by observations on the last well charted navigational mark that is seen. This is called *taking the departure.*

TAKING THE DEPARTURE. This departure must not be confused with departure used in finding the difference of longitude. *Taking the departure* consists of obtaining a good "fix," that is an accurately plotted position, from which future positions by dead reckoning are computed. It is generally done by bearings on a well charted object. There are two methods of using this departure. By the first method the vessel's position is plotted on the chart, by any of the methods described in Chapter 3, and the latitude and longitude of this position, taken from the chart, are used as the departure for future reckoning. In the second method, the bearing and distance are found of the object used for departure. The latitude and longitude of the object are taken as the point of departure for future reckoning, and the course and distance *from* the object to the vessel (the reverse of its bearing) at this time are entered as the first course and distance in the dead reckoning columns. The course from the object to the vessel is the *reverse* of the bearing of the object from the vessel.

To make this clear, suppose bearings are taken on a light and plotted on the chart, and that the position thus obtained is Lat 41° 16′N, Long 70° 50′W. This would be the point of departure according to the first method and dead reckoning positions would be computed from this. If, however, a light whose position were Lat 41° 20′N, Long 70° 50′W, bore North true, Dist 4 miles, then the position of the light can be taken as the point of departure, and the course South (reverse of bearing) true, and the Dist 4, are entered in the dead reckoning columns ahead of the first course and distance run by the vessel. In either case the result is the same.

Current. The Set of a Current is the direction *toward* which it is moving.

The Drift of a Current is the speed per hour with which it moves.

When a vessel is sailing in a current, the set and drift of which can be determined with any accuracy, as in the Gulf Stream, allowance for the current, when figuring dead reckoning, should be made as follows:

Enter the *set* of the current in the column of

courses. Multiply the *drift* of the current by the number of hours run, and enter this opposite the set in the column of distances. The effect of the current on a vessel is the same as though the vessel actually sailed the *set* and *drift*.

Summary. To sum up the subject of dead reckoning, a problem is given herewith that illustrates departure by the second method, traverse sailing, compass error, and allowance for current.

Example. Took departure at 8 A. M. on Cape Charles Lightship (Lat 37° 05′N, Long 75° 43′W) bearing (p.s.c.—per standard compass) 271°, Dist 8 miles, ship's head East (p.s.c.), var. —5°, dev. —4° (this gives true bearing of light ship 262°). Thence sailed until 10 P.M. on courses as follows.

Course	(p.s.c.)	110°,	var.	—5°,	dev.	—6°,	distance	6	miles
"	"	84°,	"	—5°,	"	—4°,	"	8	"
"	"	79°,	"	—5°,	"	—4°,	"	10	"
	"	67°,	"	—5°,	"	—3°,	"	15	"
	"	20°,	"	—6°,	"	+1°,	"	40	"
	"	286°,	"	—6°,	"	+3°,	"	48.5	"
"	"	240°,	"	—7°,	"	+8°,	"	5	"

The known current is estimated at *set* 48 (true), *drift* .5 knot per hour. Find Lat and Long at 10 P. M. by dead reckoning.

Solution: The current for 14 hours at .5 knot per hour is 7 miles, total. (See next page for table).

Lat left	......37°—05′ N	Long left	....75°—43′ W
DL 65.2′	...... 1°—05.2′N	D Lo	.... 9.8′E
Lat 10 P. M.	..38°—10.2′N	Long 10 P. M.	.75°—33.2′W

Day's Run. It is customary to calculate the total run for the preceding 24 hours every day

Course p.s.c.	*Var.*	*Dev.*	*Compass Error*	*True Course*	*Distance*	*N*	*S*	*E*	*W*
(Bearing of Lightship)				82°	8	1.1		7.9	
110°	—5°	—6°	—11°	99°	6		0.9	5.9	
84°	—5°	—4°	— 9°	75°	8	2.1		7.7	
79°	—5°	—4°	— 9°	70°	10	3.4		9.4	
67°	—5°	—3°	— 8°	59°	15	7.7		12.9	
20°	—6°	+1°	— 5°	15°	40	38.6		10.4	
286°	—6°	+3°	— 3°	283°	48.5	10.9			47.2
240°	—7°	+8°	+ 1°	241°	5		2.4		4.4
(Current)				48°	7	4.7		5.2	
						68.5	3.3	59.4	51.6
						3.3		51.6	
						65.2N = DL		7.8E = Dep	

at noon. Having the vessel's positions at two succeeding noons the problem resolves itself into finding the course and distance made good between the two positions.

Example. The position of a vessel at noon of July 12, 1914, is Lat 35°—10′N, Long 134°—01′W, on July 13, 1914, the vessel's position is Lat 36°—03′N, Long 131°—14′W. Find the course and distance made good.

Pos. July 13....36°—03′N ....131°—14′W
Pos. July 12....35°—10′N ....134°—01′W

Run .... 53′N ..D Lo 2°—47′E = 167′E

Middle Lat = 36° approximately, D Lo = 167′. From Table 2, Dep = 135.1. Entering Table 2 with Lat 53 (which is D L above), and Dep = 135.1, we obtain

Course 68½°——Distance 146 miles, made good.

Graphic Solution of Dead Reckoning. A much quicker solution of a vessel's run by dead reckoning can be made graphically on the chart. From the point of departure draw the true course on the chart. Measure from the point of departure the distance run on this course. From this second point draw the second course on the chart and lay off from this second point the distance run on the second course. By continuing this process the vessel's position by dead reckoning at any instant can be obtained. This affords an excellent check to the computations.

PART II

SEAMANSHIP

CHAPTER V

SOUNDINGS, TIDES, ETC.

APPROACHING LAND

SOUNDINGS are taken for two general purposes: first, when in shallow water, to ascertain that there is sufficient depth of water for the immediate movement of the ship, and to check the depths as given on the chart; second, to verify dead reckoning positions when on soundings in a fog or when land is not in sight.

The best aids to navigation, when running in a fog, are the sounding machine and the hand lead, and the navigator should make every possible use of them. Even in clear weather the sounding machine, or deep sea lead in lieu thereof, may be of great aid to the navigator in verifying his position. This is especially true when making a landfall. The lead and line, and the deep sea lead and line are described in Chapter 2.

THE SOUNDING MACHINE.[1] This machine pos-

[1] Whereas, a Sounding Machine will only be found on a large yacht navigated by a licensed Master, it is still considered of sufficient importance to merit a short description.

sesses advantages over the deep sea lead, for which it is a substitute, in that soundings can be obtained at great depths and with accuracy and rapidity without stopping the ship. It consists of a stand on which is mounted a reel which holds the sounding wire. Crank handles are fitted to the reel for reeling in the wire after a sounding has been taken, and a suitable brake controls the reel when the wire is running out to take a sounding. The lead is secured to the outer end of the wire. Its base is hollow to receive tallow for arming. Attached to the sounding wire, just above the lead, is the depth registering instrument enclosed in a hollow cylindrical case. Various registering devices are in use, but all depend upon the increasing pressure of water at increasing depth.

The Lord Kelvin Machine employs, for its registering device, a slender glass tube, sealed at one end and open at the other. This is coated inside with a chemical preparation which changes color on contact with sea water. This tube is placed, closed end up, in the metal container. When taking a sounding, as the lead sinks, taking the registering device with it, the air contained in the glass tube is compressed with a force dependent upon the depth. Salt water enters the open end as the air is compressed, and this makes a clearly defined line of discoloration a distance from the open end dependent on the depth. A scale is provided upon which the depth can be measured by this mark of discoloration.

Ground glass tubes may be substituted for the chemically prepared ones. When a ground glass tube is wet, it shows clear over the wetted surface. Such tubes can be used an indefinite number of times, if thoroughly dried each time.

Mechanical depth recorders can be substituted for the glass tubes. In such a device water pressure acts upon a piston against the pressure of a graduated spring, and the depth is recorded on a scale by an index pointer that is moved by the piston.

When Making Land in a Fog the sounding machine, or deep sea lead, must be kept going at half hour intervals for some hours before it is expected that soundings can be obtained. Several soundings at irregular intervals are worse than useless as they give no definite information and may lead to disaster. In using the sounding machine, be careful not to invert the tube when withdrawing it from the tube case, as that would cause water to run toward the closed end of the tube, causing a discoloration and hence a false reading. The lead must be freshly armed at each cast. Having picked up the bottom, the navigator can proceed as described under "*Piloting by Soundings in a Fog.*"

Keep a sharp lookout for any landmarks that may appear during a momentary lifting of the fog, and listen carefully for signals. If a fog signal is heard, the landmark where it is situated can be determined by reference to the Light List, which gives the characteristics of all fog signals (viz: the number and duration of the blasts). If approaching land and the soundings indicate a dangerous proximity to land, if no signals have been heard that will further aid navigation, the only safe course is to anchor or stand off shore. When running slowly in a fog (as the law requires) it must be borne in mind that the relative effect of current is increased. Sometimes, when approaching a bold bluff shore, a vessel may be warned of its proximity by having its own fog signals echoed back from the cliff. In some inland waters where cliffs are frequent, navigators depend upon this to a great extent.

Piloting by Soundings in a Fog. Soundings taken in a fog serve a much more important function than merely to give the depth of water at any one position. The vessel's position can often be found by a *series* of soundings. In thick weather, when approaching or running close to land, or in inland waters, soundings should be taken *continu-*

ously and at *regular intervals*, and the character of bottom should be noted. By laying off the soundings on tracing paper along a line that represents the track of the ship and to a scale (distances between soundings) corresponding to the scale of the chart and then moving the tracing paper on the chart so that the courses plotted are parallel to corresponding directions on the chart until the observed soundings agree with the chart soundings, the vessel's position can generally be well determined. While some waters by the quick changes in soundings along the bottom adapt themselves more readily than others to this method, there are few places where the navigator cannot at least keep out of danger by these indications. When the navigator can no longer determine with some degree of accuracy that his course leads clear of danger, it is time to anchor until more favorable conditions present themselves.

To illustrate the above by a simple example: Suppose the vessel is making 12 knots *over the ground*, and that she has steered East for 30 minutes and then NE for 30 minutes. Suppose that soundings have been taken at five minute intervals as follows: 5, 7, 8, 7, 12, 8, 10 (when course was changed), 9, 7, 6, 8, 5, 10, fathoms. On tracing paper lay off, as in Figure 21, to the scale of the chart, AB = 6 miles, BC = 6 miles (30 minute runs at 12 knots), and ABC = 135° (angle between courses). This represents the track of the vessel. Now divide AB and BC each into 6 equal parts. These are one mile intervals and the division points represent the positions at which the soundings were taken. Mark the soundings at the proper division points, and move the tracing paper over the chart, always keeping the line AB in the East and West line of the chart. At some portion of the chart it will be found that the depths on the tracing paper correspond to depths on the chart (if the soundings have been carefully taken) and the vessel's position is at the point of the last sounding.

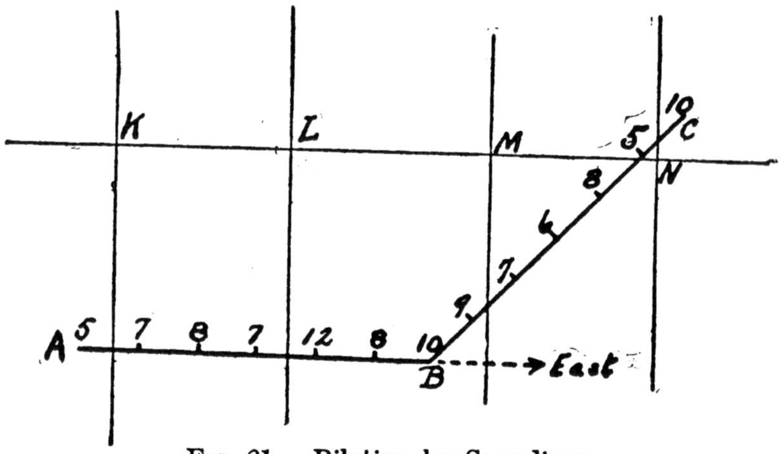

FIG. 21.—Piloting by Soundings.

Orienting the tracing paper is facilitated by drawing a few meridians, K, L, M, N, thereon. If there is much range to the tide the stage of the tide must be noted when soundings are taken and allowance must be made for the height of tide. The soundings on the chart are for mean low water.

EFFECT OF WIND AND BAROMETER ON SOUNDINGS. When navigating waters where the depth exceeds the vessel's draft by only a small amount, it must be borne in mind that a strong wind or unusually high barometer may cause the water at low tide to fall below the depth indicated on the chart.

PILOTING AMONG CORAL REEFS. When piloting among coral reefs or banks, a time should be chosen when the sun is astern. Conning should be done from an elevated position forward. When the observer is high up, the line of demarcation between a reef or shoal and deep water is very clearly seen.

When passing *between* shoals or dangers where there are no well charted navigational marks a mid-channel course should be steered. Too much emphasis cannot be laid upon this point. Do not save seconds by passing close to a danger when a safe course offers. Steering a mid-channel course by eye is a simple matter, as the eye can make a close estimate in a case of this kind.

TIDES

To an observer tides present themselves in two different manners, by alternate elevation and depression of the water level, and by alternate inflows and outflows of streams. Properly speaking *tides* should refer to the vertical motion and *tidal currents* to the horizontal flows. However, popular usage ascribes both these meanings to *tides.*

When the water has reached a maximum level it is called *high tide,* or *high water.* When it has reached a minimum level it is called *low tide,* or *low water.* The interval at high and low water, when there is no perceptible movement, is called the stand.

Between low and high water, when the tide is setting from the sea toward the land, the horizontal movement is called *flood tide.* When the horizontal movement sets from the land toward the sea, between high and low water, it is called *ebb tide.* The interval of change between flood and ebb, when there is no perceptible horizontal movement, is called the *slack.*

The Cause of the Tides is the difference in the attraction of the moon (and to a less degree of the sun) upon the various unit masses of water and the attraction of the moon on the earth itself. This difference of attraction combined with the relative periodic movements of the moon and earth produce the periodic ocean disturbances known as tidal phenomena.

Establishment. The moon being the dominant factor in tide production, it is apparent that the phenomena should bear some relation to the lunar month (28 days). High and low water occur, on the average of these 28 days, at about the same intervals *after the transit* of the moon over the meridian. These nearly constant intervals, expressed in hours and minutes, are known respectively as the *high water lunitidal interval* and the *low water lunitidal interval.* The interval between the moon's meridian passage at any place and the time of the next succeeding high water, as observed on the days when the moon is at full and change (new), is called the *establishment;* it is also spoken of as the *time of high water on full and change days* (abbreviated "H. W. F. & C."), for since the moon's meridian passage on these days occurs at midnight and noon (of the lunar day), the establishment is approximately the time of high water. If to the time of high water we add or subtract 6 hours 31 minutes (¼ of a lunar day) the result will be the time of low water.

Range, Spring and Neap Tides. The *range* of a tide is the difference in height between high and low water. At new and full moon the sun and moon produce high tides at the same times. The effect of the sun augments that of the moon and the resultant high tides are higher than the average. Similarly the low tides are lower than the average, and the range of tides is greater. Tides at this season are called *spring tides*. At the time of first and third quarters of the moon the high tides due to the moon occur simultaneously with the low tides due to the sun. The resultant high and low tides are less than the average and the range is at its minimum. Tides occurring at this season are called *neap tides*. Tidal currents (*flood* and *ebb*) are strongest at spring tides and weakest at neap tides.

Tidal Currents. It must be remembered that the periods of *flood* and *ebb* in any locality are not necessarily coincident with the periods of rise and fall of the tide. The inward set of the surface current does not always cease when the water has attained its maximum height. Local conditions may be such that flood may continue after high water has been reached, or vice versa.

This may be more apparent by comparing two tidal basins, one having a large open entrance and the other having a narrow restricted opening. In the first case the process of filling and emptying the basin keeps pace with the outside sea level and ebb and fall, flood and rise, occur at practically the same time. In the second case the restricted entrance retards filling the basin so that the height of water without may reach a maximum long before the basin fills. In this case *flood continues*, possibly hours, *after high water occurs*, and in a like manner the restriction will cause ebb to continne long after low water has occurred outside.

Times of High and Low Water. The simplest and quickest method of obtaining the times of high and low water and other tidal data is by use of a Tide Table. One is published by the U. S. Coast and Geodetic Survey annually, and gives the times of high and low water at many seaports. From these others may be deduced. Much other tidal data is given in this publication. The daily papers of many marine ports give the times of the tides for several days ahead of the date of issue.

When no tidal table or data is obtainable the time of high water can be computed by use of the Nautical Almanac as follows: To the time of the moon's meridian passage add the lunitidal interval (H. W. F. & C.), obtained from the chart or from Bowditch's Useful Tables.

The time of the moon's meridian passage is obtained as follows: From the Nautical Almanac pick out the Greenwich Mean Time of the Upper Transit of the Moon at Greenwich for the local date (Page IV). Also pick out the daily variation on the same line. With this *variation* and with the *Longitude* as arguments enter table 11, Bowditch, and obtain the correction to be applied to the G. M. T. of Greenwich Transit. The result is the Local Mean Time of the moon's meridian passage. Now add to this the time of H. W. F. & C. and the result is the Local Mean Time of high water. This can be converted to Standard time by applying the difference in time between the longitude of the place and the Standard Meridian.

Appendix IV, Bowditch, contains the mean lunitidal interval of high and low water for many places. The charts give the establishments of many ports with sufficient accuracy for this work.

General. To sum up, when approaching land or harbor, the navigator must know the draft of the vessel. He must make himself familiar with every detail of the charts he will use, and must form a mental picture of the land and aids to navigation that he will sight. Allowance must be made for the effect of the position of the sun or moon on the appearance of objects sighted. He must be familiar with the characteristics of all lights, buoys, fog signals, and other aids to navigation that he will use, and with the state of the tide and currents in channels he will navigate. He should select beforehand the objects that he will use for bearings. He should carefully check all buoys to prevent confusion. Ranges should be selected and lines drawn to indicate safe courses and danger bearings where possible. The track of a vessel entering port should be laid down on a chart before entering, and this should be carefully inspected to see that it leads clear of all possible danger.

The vessel's position must be frequently plotted on the chart and should never be in doubt for an instant. Soundings should always be taken when on soundings, whether the weather be clear or cloudy. The navigator should familiarize himself with the Inland Rules of the Road given in Chapter 8 before entering pilot waters.

CHAPTER VI

LIGHT AND BUOY SYSTEM OF THE UNITED STATES

LIGHTS

LIGHTS are distributed along the coast, at harbor entrances, in harbors, and at other points to aid navigation. They may be placed in lighthouses, on lightships, or on buoys. When in lighthouses, there is generally one light to a house; when on lightships there is generally more than one.

Lights are classified as fixed, flashing, intermittent, revolving, and fixed and flashing. This is necessary so that when a light is sighted it can be identified. It is obvious that if all lights were the same, the navigator might become confused when approaching land if his position were at all uncertain.

A Fixed Light is one that shows uninterruptedly at all times.

A Flashing Light is one that shows a short flash and is then occulted for a long interval.

An Intermittent Light is one that shows a long flash and is occulted for a period shorter than the flash.

A Revolving Light is one that gradually increases in intensity, then gradually decreases in intensity until it is occulted, and then gradually increases again.

A Fixed and Flashing Light is a combination of the first two.

Lights may be red or white in color. As stated before, the lights are given different characteristics so that they may be readily distinguishable. Flashing lights sometimes flash a number for this purpose. Generally speaking, main coast lights are white, although there are exceptions to this rule. Harbor lights may be white or red. Red lights are used to mark dangers, such as ends of breakwaters, etc.

Red Sectors. Many white lights have red sectors, that is, the light shows white over part of the horizon, and red over other parts. Red sectors are used to mark dangers. In this case the light shows white as long as the observer is in safe navigable waters, but when in the same sector as a shoal or other danger the light shows red. In this case the vessel is safe as long as it is in the white sector.

Light Lists. The United States is divided into a number of lighthouse districts. A list of lights is published by the Hydrographic Office for each district, and is sent on request to any ship cap-tain. These lists include all authorized lights, with their description, etc. Lights in lighthouses are described in the light lists as follows: (An example is taken from a list.)

"Lighthouse color, white; foundation brown; lantern yellow. (This is to identify it by day.)

"Hexagonal screw-pile structure; light 44 feet above high water.

"Bearings. (Here its bearings from other well charted objects are given.)

"Character, 3000 candlepower, red, visible 8½ miles (for a height of eye of 15 feet above the sea level).

"Fog Signal. (Here a description of the fog signal installed at the light is given.)"

Light Vessels are moored at sea outside of important harbors, and on the edge of important shoals on the coast where it is impracticable to plant lighthouses. They are described in the same light list. The names of the light vessels are painted on their sides. These names are taken from the shoal that the light guards, or from other sources. Numbers are often used in lieu of names. The description of a light vessel from the light list is given as follows:

"Description, white; masts yellow; topmast and day marks black.

"Rig. 2 masts, schooner rigged, oval day marks at each mast head.

"Bearings. (Here follow the bearings of the light from well charted objects.)

"Fog Signal, on whistle, blast of 3 seconds, silent 60 seconds, blast of 3 seconds, silent 60 seconds, etc.

"If the whistle is out of order the signal is made on the bell as follows: 3 strokes, silent 60 seconds, 3 strokes, silent 60 seconds, etc."

Light Buoys are used to mark the entrance to harbors, to guard shoals, and to mark the main channel of large harbors. They may be fixed or flashing, and red or white; all light buoys are described in the light lists.

BUOYS

All buoys, together with their location, are described in the light and buoy list of each district. Certain rules govern buoys which, if remembered, make navigation by buoys a simple matter.

The particular feature about a buoy is its color. All buoys of each marked channel are numbered from seaward.

Red buoys have even numbers and must be left on the starboard hand when entering harbor.

Black buoys have odd numbers and must be left on the port hand when entering harbor.

When a channel has two entrances from seaward local rules must govern. Thus on the Maine coast where channels have two sea entrances buoys in thoroughfares and passages are numbered and colored for entering from Eastward.

Buoys painted with red and black horizontal stripes mark shoals or other dangers, and should be given a wide berth. Channel buoys are frequently anchored abreast of these to mark the channel.

Buoys painted with white and black vertical stripes are midchannel buoys and should be passed close to.

Yellow buoys mark quarantine anchorages.

White buoys mark anchorages.

Shapes of Buoys. Buoys are shaped as follows:

Can, cylindrical.

Nun, a truncated cone, or

Spar.

With the exception of spar buoys, all buoys are made of sheet iron, with water tight compartments to prevent sinking in case of damage.

Where there is more than one channel in a harbor the different buoys are used to mark different channels; nun buoys mark the main channel, can buoys, the secondary channels, and spar buoys, minor channels. When there is but one channel, nun buoys are placed on the starboard side and can buoys on the port (when entering from seaward).

Buoys that mark important shoals on the coast are marked with letters or numbers; thus the buoy that marks the Frying Pan Shoal has *F.P.* painted on it.

Bell buoys, whistling buoys, and buoys with balls, baskets, and other shapes mounted on a perch, mark turning points in the channel. The color and number of the buoy indicates on which side to pass. Where there is much ice, bell and gas buoys are frequently removed in winter, leaving only the spar buoy marking the position.

FOG SIGNALS

Fog Signals are established at all important lighthouses and light vessels to aid navigation in a fog. Each lighthouse and light vessel has a distinguishing signal so that it can be identified, and full descriptions of these are given in the light lists. The signal may be given on the whistle or bell.

CHART NOMENCLATURE

Lights are indicated on some charts by a yellow spot surrounding a black dot in the case of a white light, and a red dot in the case of a red light. Its characteristics, color, and range of visibility are marked on the chart abreast of it.

Buoys are marked on the chart with their numbers and the following abbreviations to indicate their characteristics:

B — black.
R — red.
H.S.— black and red, horizontal stripes.
V.S.— black and white, vertical stripes.
C — can.
N — nun.
S — spar.

The Character of the Sea Bottom is indicated as follows: one letter abbreviations are used to indicate kind of bottom, two letters to indicate color of bottom, and three letters for other qualifications, as follows:

M — mud.	Yl — yellow.	Brk — broken.
G — gravel.	Gy — gray.	Sml — small.

Soundings are indicated in fathoms unless otherwise noted. On harbor charts (1/80,000 scale) sounding will probably soon be changed to feet. Thus, 50 on the chart means 50 fathoms *at mean low water.* $\frac{0}{50}$ means, "no bottom at 50 fathoms."

Miscellaneous Marks. Other common indications on a chart are:

Rock awash at low water . . . . 

Rock, sunken . . . . 

Danger of doubtful existence (marked abreast) . . . . E.D.

Danger of doubtful position (marked abreast) . . . . P.D.

Anchorage, large vessels . . . . 

Anchorage, small vessels . . . . 

Wreck . . . .  or . . . . 

Lightship . . . . 

Currents are marked by an arrow, with two barbs for flood, and one barb for ebb; the number on the arrow indicates the strength in knots. If the current is tidal, one, two, or three cross bars indicate the 1st, 2d, or 3d quarter of the flow. Thus  indicates flood current in the 1st quarter, strength 3 knots per hour. Likewise  indicates ebb current in the 3d quarter, strength 5 knots.

CHAPTER VII

WEATHER

WINDS

CAUSES of Winds. Wind is air in horizontal motion. It is defined by its direction and force. The direction of the wind is the point of the compass *from* which it proceeds. Its force is generally measured by the Beaufort Scale described in Chapter II and depends upon its velocity. If air is warmer in one place than in an adjacent place, the warm air will rise and will be replaced by air flowing in from the second place. This creates a wind from the second to the first place. To take another view of the matter, in the warmer place the barometric pressure is lower than in the second (cooler) locality where the air is descending.

The direction of winds is always from a place of high barometer to one of lower pressure. The Weather Bureau supplies data from which daily weather charts are plotted, showing the distribution of barometric pressures over the United States and its adjacent waters. These charts can be consulted in any large seaport. Weather predictions made from these charts are invaluable to the mariner.

The greater the barometric range between two adjacent places, the more violent the disturbance accompanying the transfer of air from the region of high barometer (called a "high") to the region of lower pressure (called a "low"). When sufficiently violent we have a gale or storm.

Ascending currents of warm air carry moisture that has evaporated from the sea. As the air ascends it encounters lower temperatures which condense the moisture. If this moist air ascends to a sufficient height, or having ascended moves to a colder region, the moisture in the air is sufficiently cooled to be precipitated and we have the phenomenon of rain.

Land and Sea Breezes. Generally speaking, the land is warmer than the adjacent sea by day, and cooler at night. This is due to the fact that the land as a whole absorbs and radiates heat more rapidly than will a large body of water; this is especially the case in summer.

As a consequence of the above, a variation of pressure between the land and sea is established, which, though small, nevertheless is sufficient to affect the local winds. Under normal conditions, the wind blows from the sea toward the land during the day, and from the land toward the sea at night.

TRADE WINDS. In the general terrestrial distribution of the atmosphere the equator is belted by a region of low pressures. To the North and South of this belt are other belts of high pressure along the latitudes of approximately 30° North and South. Consequently, winds blow rather constantly toward the equator over a considerable area. Due to the effect of the earth's rotation these winds are from the Northeast in the Northern hemisphere and from the Southeast in the Southern hemisphere. These prevailing winds in the low latitudes are called the *trades.*

THE DOLDRUMS. In the equatorial belt of low pressures there is little horizontal motion to the atmosphere. The atmosphere is slowly rising and the winds are stagnant, blowing fitfully in light airs from first one and then another point of the compass. Due to the constant evaporation and ascending air currents the weather is generally cloudy, with frequent thunder storms. This region is called the *Doldrums.*

THE HORSE LATITUDES. The belt of high barometric pressures that lies along latitudes in the Thirties, North and South, is another region of comparative calms. Here the breezes are also light, but the weather is clear in contrast with that of the doldrums. The reason lies in the fact that over this region a downward current of air prevails. This region is called the *Horse Latitudes:*

BAD WEATHER

Bad Weather is a comparative term. A heavy squall that would be considered bad for a very small boat would not inconvenience a large vessel. For this reason an attempt has been made to classify bad weather in an unscientific manner that will be intelligible to the lay mind.

No amount of rain or snow would endanger the safety of even a small boat. The factor in weather that must be considered is *Wind.* Although inconvenient and uncomfortable, rain will not affect the placidity of the sea; the condition of the sea depends upon the strength of the wind, and *the safety of a vessel depends upon the strength and direction of the wind and the state of the sea.*

For the benefit of navigators of boats and small vessels, bad weather might be classified as follows: (1) *squalls;* (2) *gales;* and (3) *storms.* This classification depends upon the strength and duration of the bad weather. *Squalls* are of short duration, the wind is variable and a good sailor is safe in almost any small boat. *Gales* are of longer duration, the wind is stronger and steadier, and the seas are higher. *Storms* are often of several days' duration, the wind follows certain laws, being of cyclonic origin, and the seas become so high as to be dangerous to any but good sized vessels.

SQUALLS may be encountered in almost any locality. At some seasons they are of daily occurrence in the tropics. There is little previous warning before a squall breaks, although a long period of squally weather will generally be preceded by a falling barometer. A squall may or may not be accompanied by rain. When accompanied by rain the warning comes as a grayish vertical curtain (the rain) obscuring *part* of the horizon. When this curtain is to windward the squall will probably strike the observer. When to leeward it may or may not be encountered. Often squalls work to windward. If a squall does not contain rain, the first indication is local agitation of the surface of the water. Wind on the water can be seen for a considerable distance. Squalls are accompanied by capricious shifts of wind of a puffy nature. The wind speed will rarely exceed 35 miles an hour.

Handling the Boat. Very small boats under sail should get the sail off before the squall strikes, as the uncertainty of the wind shifts might embarrass. Any well handled boat is safe in a squall.

If the sea becomes so high as to be uncomfortable on the course steered, run with the sea astern. If there is danger of the sea breaking over the stern, round to and lay head to the sea. The head may be kept up to the sea by putting a drag over the bow. A drag may be made by lashing together a few oars, a spar and sail, or anything that will lie on or near the surface.

GALES may be encountered in almost every locality. In our own waters they may be found in the Gulf of Mexico, along the entire Atlantic and Pacific sea coasts, and in the Atlantic along paths North of 30° Lat. and South of 25° Lat. The barometer gives early indications of a gale. A steadily falling barometer, accompanied by a steadily increasing wind indicates the approach of a gale. A dull lowering sky, Nimbus clouds or "scud," with Alto-stratus or Cirro-stratus above, and a rising sea, all precede a gale. A red sky in the morning and halos around the moon or sun may be early indications. Gales are generally accompanied by rain, hail, sleet, or snow. The wind varies from 40 to 60 miles per hour. As the wind increases in intensity the barometer falls rapidly. These two conditions taken together are almost certain signs of a gale.

Handling the Boat. Small boats should seek shelter. Sail should be doused on sailing craft.

Small craft will do well to run with the sea aft, with just enough speed on to keep the sea from breaking over the stern. If for any reason it is impossible to run this way and it becomes necessary to lay to, do so with the sea ahead, with a good drag over the bow. If necessary to lie hove to in fair sized craft it may be found that the vessel is more comfortable if hove to stern to the sea without a drag. Large vessels can generally pursue their course through a gale, but the speed should be reduced as the gale strengthens.

STORMS are invariably of cyclonic origin and follow well established rules in their movements and wind shifts. It is beyond the scope of this work to discuss the whole subject, but a few indications of their approach and rules for handling vessels in a storm will be given. The duration of a storm may be several days, and during this entire period the wind blows at 75 miles per hour or more. It gradually shifts in direction according to well established rules given hereafter. Indications of its approach are available for at least 24 hours before the storm breaks in all its fury.

Early Indications. A *storm* is preceded by an abnormal *rise* of the barometer, with cool, dry, fresh winds and a cessation or reversal of the ordinary land and sea breezes. The atmosphere becomes very transparent. A long low swell is present at a great distance from the storm center, sometimes several hundred miles, and this is occasionally accompanied by hurricane rollers. When there is no intervening island or land to divert them, the direction of the rollers indicates the direction of the storm center. Feathery Cirrus clouds form on the horizon and radiate from a point on the horizon, which point also indicates the direction of the storm center.

Unmistakable Signs. As the storm develops and comes nearer the sky becomes hazy and is covered by a veil of Cirrus clouds which form halos by day and night. The barometer falls very rapidly and becomes unsteady. The air is heavy, hot, and moist and the sky assumes red and violet tints at dawn and sunset. A low solid hurricane cloud bank forms on the horizon having the appearance of land. Squalls break off and diverge from this cloud bank and later these squalls pass the line of center of the bank. Fine misty rain forms with a heavy cross sea. As the storm develops the wind rises and the barometer falls more rapidly and becomes more unsteady. The wind blows at more than 90 miles per hour.

Location of Storm Center. During the early observations the location of the storm center can be found by the directions of the rollers or of the storm bank as given above. As the storm develops and these can no longer be observed, the direction of the center can be taken as 10 to 12 compass points from the direction of the wind, to the *right in the Northern hemisphere* and to the *left in the Southern hemisphere.* As the storm increases in intensity and the center approaches, after the barometer has fallen as much as one-half inch, the direction of the center may be taken at 8 points from the direction of the wind.

Wind Shifts. In the Northern hemisphere the wind rotates around the storm center in an anticlockwise direction. To an observer on a vessel if the wind appears to shift to the right, the vessel is in the dangerous zone of the storm; if the wind appears to shift to the left the vessel is in the navigable zone; if the wind blows steadily from one direction as it increases in intensity, the vessel is in the path of the storm center.

Handling the Boat. Small craft must seek shelter on the approach of a storm. Only large vessels are safe in them. There is apparently no limit to the degree of intensity of the wind, and the seas become very high and irregular. If forced to lie to this may be done with a drag, but a safer way is to run before the storm at just sufficient speed to keep the seas from breaking over the quarter. The following table shows the maneuvers for sailing vessels caught in a storm:

IN THE NORTHERN HEMISPHERE

Heave to on the Starboard Tack to Observe the Wind

Wind	*Zone*	.
If the wind hauls to the right	The vessel is in the right or dangerous zone	Run close hauled on the starboard tack If obliged to lie to do so on the starboard tack.
If the wind hauls to the left	The vessel is in the left or navigable zone	Run with the wind on the starboard quarter If obliged to lie to do so on the port tack
If the wind remains steady	The vessel is in the path of the storm center	Get the wind on the starboard quarter and keep that compass course If obliged to lie to do so on tack that the wind and sea will draw aft

Large, full powered vessels will do well in the dangerous zone to run with the wind ahead or on the starboard bow as long as possible. This will work the vessel away from the storm center. The speed must be reduced, in fact just make steerage way. When it is no longer practicable to run into the seas lie to or run with the wind on the starboard quarter with just sufficient speed to keep the seas from breaking aboard.

When in the left or navigable zone, run with the wind on the starboard quarter at a reduced speed. Lie to when necessary. If the vessel is in the track of the storm center get the wind on the starboard quarter, note the compass course, and keep this course until the vessel has worked its way out of the storm.

Tornadoes and Water Spouts. A tornado might be likened to a concentrated storm, although this would not be technically correct. It depends upon an unstable and very humid state of the atmosphere, and is cyclonic in nature. A tornado may only be a few hundred yards in diameter. During an unstable and very humid state of the atmosphere, a warm, moist air current, stronger than usual forces its way up, and once started, increases in violence.

The upward velocity and velocity of gyration are extremely high, the former reaching as much as 150 miles per hour, and the latter as much as 200 miles per hour. As the diameter is very small the vortex is very steep, and the barometer may fall from the normal to nearly zero, a state of perfect vacuum. This accounts for the great damage done by this class of storm, the inside pressure of a building being normal and the outside pressure being nearly zero, the tendency is for the building to burst. This class of storm, due to its causes of formation, is always accompanied by rain around its outside. There is no rain in its center, the upward tendency of the atmosphere preventing the moisture from descending.

A Water Spout is simply a tornado formed at sea in atmosphere laden with moisture where the dew point stratum is comparatively near the earth's surface. It is an erroneous belief that the water column is formed by water sucked up from the sea. This is not a fact. The water column is due to moisture drawn down from the heavy moisture-laden atmosphere. It is doubtful whether water is drawn up from the sea to any great extent, and it is certain that the sea surface is not drawn up more than 30 feet, the height of the water barometer.

CHAPTER VIII

RULES OF THE ROAD

SHIPPING, both on the high seas and in pilot waters, is bound by certain rules as to the lights to be carried, navigational signals to be made, and the manner in which to maneuver to avoid collision.

International Rules of the Road govern these matters on the high seas and they are enforced by Maritime Law. They are agreed to by the principal nations and are promulgated by the legislative bodies thereof. These rules always apply when on the high seas or in waters not under the territorial jurisdiction of any particular nation.

Inland Rules of the Road. Rules for the conduct of vessels in the territorial waters of different countries are promulgated by the legislative bodies thereof. Those promulgated by the U. S. Congress are similar in many respects to the International Rules. Important differences therein are pointed out in this chapter. These rules are not applicable to vessels on the Great Lakes, where a special set of rules govern.

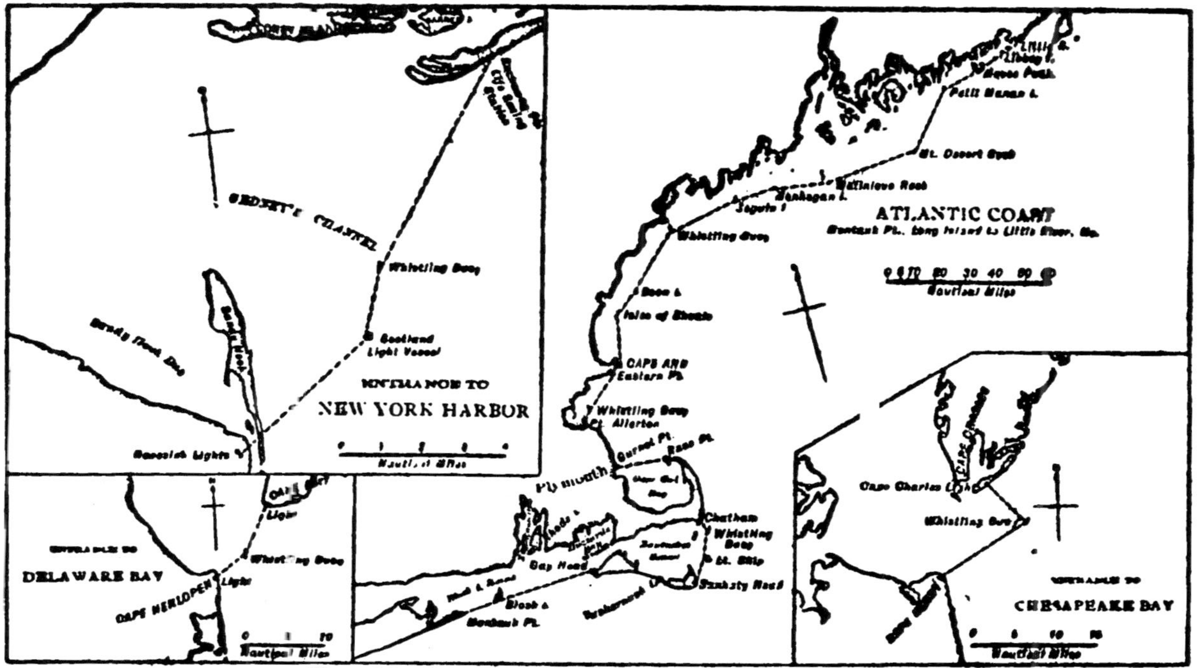

FIG. 22. INLAND RULES OF THE ROAD

-------------This dotted line represents demarcation between Inland and International Rules.

These rules apply to all vessels plying the inland waters of the United States; the limits of these waters are shown in Figure 22. All waters inside the dotted lines shown in this Figure are considered U. S. Inland waters for navigation. The rules are too voluminous to quote *in toto* so a summary of the important ones follows:

INLAND RULES OF THE ROAD

Classification of Vessel. Any vessel propelled *by machinery* is classed as a *steamer.* Steamers when under sail and not under steam are classed as sailing vessels. In this case they must carry, by day, a black ball or shape forward to distinguish them from steamers under way.

A vessel is considered *underway* when she is not at anchor, or made fast to the shore or aground.

Lights of Vessels. *All lights that are required by the rules of the road must be shown from sunset to sunrise, in all weathers, and during such time no other lights that may be mistaken for the prescribed lights shall be exhibited.*

1. Steamer Lights When Underway. (a) Forward, a white light, visible at least 5 miles over an arc of 20 points of the horizon, from ahead to 2 points abaft both beams. This is known as the *masthead light.*

(b) On the starboard side, a green light, visible at least 2 miles over an arc of 10 points, from ahead to 2 points abaft the starboard beam. This is known as the *starboard side light.*

(c) On the port side, a red light, visible at least 2 miles over an arc of 10 points, from ahead to 2 points abaft the port beam. This is known as the *port side light.*

(d) An additional light similar to that described in (a) shall be carried aft and higher than (a). This additional light forms, with the masthead light, range lights.

(e) The green and red lights must have inboard screens so placed as to prevent them showing across the bow. These lights are carried lower than the masthead light.

The additional light described in (d), which with the masthead light forms a range, is compulsory in Inland Waters, except in the case of seagoing vessels and ferry boats. The International Rules make this light optional outside Inland Waters.

It is apparent that when a vessel on an even keel, carrying range lights, is seen head on, the lights are seen one above the other; if the vessel changes course the lights will open out, the lower one away from the upper one in the direction to which the vessel's head is changing. Such change gives instant notice of change of course and is more reliable than side lights because the distance at which the range lights can be seen is so much greater.

Care should be exercised not to confuse range lights with towing lights, given under 2 (a).

2. Special Steamer Lights. (a) *When towing*, a steamer carries its regular lights and an additional light similar to the masthead light in a vertical line therewith, and, if towing more than one vessel and the tow exceeds 600 feet, it carries two such additional lights. These lights must be carried at least 3 feet apart.

International Rules require that outside of inland waters towing lights must be at least 6 feet apart. Care must be taken not to confuse towing lights with range lights. When range lights and towing lights are both carried, four lights in a vertical line may be seen. At distances beyond four miles 2 lights 6 feet apart blend into one.

(b) *Pilot Vessels* when on their station shall carry forward a white light visible all round the horizon and shall exhibit a flare-up light at short intervals not to exceed 15 minutes.

On the near approach to or of other vessels they shall flash their side lights at short intervals to indicate their heading.

In addition to the above special lights required by the Inland Rules, the International Rules require the following lights, outside inland waters:

(c) *A Vessel Not Under Command* carries forward two red lights in a vertical line one over the other, visible all round the horizon at least 2 miles. In the daytime it carries two black balls or shapes.

(d) *A Vessel Engaged in Laying or Picking Up a Cable* carries forward three lights in a vertical line. The highest and lowest lights are red and the middle light is white. All are visible all round the horizon at least 2 miles.

(e) When making way through the water Towing Steamers, Cable Vessels, and Vessels Not Under Command must carry side lights as prescribed for steamers.

3. SAILING VESSEL LIGHTS. *A sailing vessel or a vessel being towed* must carry the side lights prescribed for steamers but must *not* carry a masthead light.

4. SMALL STEAMERS, SAIL VESSELS, AND ROW BOATS. (a) *Steam vessels of less than 40 tons* carry a masthead light visible 2 miles and side lights visible 1 mile, but in lieu of side lights *may* carry a combined lantern showing a green and red light from right ahead to 2 points abaft the beam on their respective sides. This lantern is carried at least 3 feet below the masthead light.

(b) *Rowing Boats*, whether under oars or sail, shall have ready at hand a white lantern which shall be temporarily exhibited in time to prevent collision.

5. LIGHTS FOR AN OVERTAKEN VESSEL. A vessel which is being overtaken shall show from her stern a white light or flare up light to the overtaking vessel.

6. ANCHOR LIGHTS.[1] (a) A vessel under 150 feet in length shall carry forward at a height not over 20 feet a white light visible all round the horizon at least 1 mile.

(b) A vessel of 150 feet or over shall carry forward at a height between 20 and 40 feet a white light visible all round at least 1 mile. At the stern she shall carry a similar light at least 15 feet lower than the forward light.

In addition to the above the International rules prescribe:

(c) *A Vessel Aground* in or near a fairway shall carry anchor lights as above and in addition the two red lights prescribed in 2 (c). No provision is made for vessels aground in the Inland Rules and they carry the regular anchor lights.

7. LIGHTS FOR FERRY BOATS are prescribed by special rules established by the Supervising Inspector General of Steam Vessels and approved by the Secretary of Commerce.

[1] No distinction is made between steamers and sailing vessels in anchor lights.

8. Rafts and Other Water Craft not specifically provided for, navigating by hand power, horse power, or by the current of a river shall carry one or more good white lights in such manner as is provided by the Board of Supervising Inspectors of Steam Vessels.

SOUND SIGNALS

Sound Signals for Passing Steamers. *A Short Blast* is one of about one second duration.

When steamers are in sight of one another change of course is indicated by the following signals:

(a) One short blast means "I am directing my course to starboard."

(b) Two short blasts means "I am directing my course to port."

(c) Three short blasts means "My engines are going full speed astern."

Sound Signals for Fog. Fog signals are given by steamers on the whistle or siren.

Fog signals are given by sailing vessels and vessels being towed on the fog horn.

A Prolonged Blast means one of 4 to 6 minutes' duration.

1. Fog signals must always be made by all vessels in fog, mist, falling snow, or heavy rain storm, whether by day or night.

2. *Steam Vessels Underway.* A steam vessel underway shall sound at intervals of not more than one minute, a prolonged blast.

The International Rules differ materially from the above and are quoted here for use on the high seas:

(a) A steam vessel *having way upon her* shall sound, at intervals of not more than 2 minutes, a prolonged blast.

(b) A steam vessel *underway, but stopped, and having no way upon her*, shall sound at intervals of not more than 2 minutes, two prolonged blasts, with an interval of about one second between.

The navigator must hold clearly in his mind the fact that the "sound signals in a fog" must be used at all times in inclement weather, *up to the moment of sighting another vessel.* When two vessels come in sight of each other then "sound signals for passing vessels" should be used. It is a common error among seamen to use passing signals when two vessels are within *sound but not sight* of each other. This practice cannot be too severely condemned.

3. *Sailing Vessels Underway* shall sound at intervals of not more than 1 minute, when on the starboard tack, one blast; when on the port tack,

two blasts; when with the wind abaft the beam, three blasts in succession.

4. *Vessels at Anchor* shall, at intervals of not more than one minute, ring the bell rapidly for about 5 seconds.

5. *A Vessel When Towing* or being towed shall sound at intervals of not more than 2 minutes, three blasts in succession, namely: One prolonged blast followed by two short ones.

The International Rules prescribe this same signal for vessels employed in laying or picking up a cable and for vessels underway but not under command.

6. Rafts and other small craft not specifically provided for navigating by hand power, horse power, or by the current of a river, shall sound a blast of the fog horn or equivalent signal, at intervals of not more than one minute.

SPEED IN A FOG

Every vessel shall, in a fog, mist, falling snow, or heavy rain storms, go at a moderate speed, having careful regard to the existing circumstances and conditions.

A steam vessel hearing, *apparently forward of her beam,* the fog signal of a vessel the position of which is not ascertained shall, so far as the circumstances of the case admit, stop her engines, and then navigate with caution until danger of collision is over.

What constitutes moderate speed in a fog is a much mooted question. However, it can be defined as such speed as will with certainty prevent collision. Clearly speed that might be permissible on the high seas would be too high for crowded inland waters. It might even be necessary to stop and anchor in some cases.

STEERING AND SAILING RULES

1. Preliminary. Risk of collision can, when circumstances permit, be ascertained by carefully watching the compass bearing of an approaching vessel. *If the bearing does not appreciably change*, such risk should be deemed to exist.

2. Sailing Vessels. When two sailing vessels are approaching each other, so as to involve risk of collision, one of them will keep out of the way of the other, as follows:

(a) A vessel which is running free shall keep out of the way of a vessel which is closehauled.

(b) A vessel which is closehauled on the port tack shall keep out of the way of a vessel which is closehauled on the starboard tack.

(c) When both are running free, with the wind on different sides, the vessel which has the wind on the port side shall keep out of the way of the other.

(d) When both are running free, with the wind on the same side, the vessel which is to windward shall keep out of the way of the vessel which is to leeward.

(e) A vessel which has the wind aft shall keep out of the way of the other vessel.

3. Steam Vessels. When steam vessels are approaching each other head and head, that is end on, or nearly so, it shall be the duty of each to pass on the port side of the other; and either vessel shall give, as a signal of her intention, one short and distinct blast of her whistle, which the other vessel shall answer promptly by a similar blast of her whistle, and thereupon such vessels pass upon the port side of each other. But if the courses of such vessels are so far on the starboard of each other as not to be considered as meeting head and head, either vessel shall immediately give two short and distinct blasts of her whistle, which the other vessel shall answer promptly by two similar blasts of her whistle, and they shall pass on the starboard side of each other.

The foregoing only applies to cases where vessels are meeting end on, or nearly end on, in such a manner as to involve risk of collision; in other words, to cases in which, by day, each vessel sees the masts of the other in line, or nearly in line, with her own, and by night to cases in which each vessel is in such a position as to see both the side lights of the other.

It does not apply by day to cases in which a vessel sees another ahead crossing her own course, or by night to cases where the red light of one vessel is opposed to the red light of the other, or where the green light of one vessel is opposed to the green light of the other, or where a red light without a green light or a green light without a red light is seen ahead, or where both green and red lights are seen anywhere but ahead.

(b) If, when steam vessels are approaching each other, either vessel fails to understand the course or intention of the other, from any cause, the vessel so in doubt shall immediately signify the same by giving several short and rapid blasts, not less than four, of the steam whistle. This is commonly known as the danger signal.

(c) Whenever a steam vessel is nearing a short bend or curve in the channel, where, from the height of the bank or other cause, a steam vessel approaching from the opposite direction can not be seen for a distance of half a mile, such steam vessel, when she shall have arrived within half a mile of such curve or bend, shall give a signal by one long blast of the steam whistle, which signal shall be answered by a similar blast given by any approaching steam vessel that may be within hearing. Should such signal be so answered by a steam vessel upon the farther side of such bend, then the usual signals for meeting and passing shall immediately be given and answered; but if the first alarm signal of such vessel be not answered, she is to consider the channel clear, and govern herself accordingly.

When steam vessels are moved from their docks or berths, and other boats are liable to pass from any direction toward them, they shall give the same signal as in the case of vessels meeting at a bend, but immediately after clearing the berths so as to be fully in sight, they shall be governed by the steering and sailing rules.

(d) When steam vessels are running in the same direction, and the vessel which is astern shall desire to pass on the right or starboard hand of the vessel ahead, she shall give one short blast on the steam whistle as a signal of such desire; and if the vessel ahead answers with one blast, she shall put her helm to port; or if she shall desire to pass on the left or port side of the vessel ahead, she shall give two short blasts of the steam whistle as a signal of such desire; and if the vessel ahead answers with two blasts, shall put her helm to starboard; or if the vessel ahead does not think it safe for the vessel astern to attempt to pass at that point, she shall immediately signify the same by giving several short and rapid blasts of the steam whistle, not less than four, and under no circumstances shall the vessel astern attempt to pass the vessel ahead until such time as they have reached a point where it can be safely done, when said vessel ahead shall signify her willingness by blowing the proper signals. The vessel ahead shall in no case attempt to cross the bow or crowd upon the course of the passing vessel.

(e) The whistle signals provided in the rules under this article for steam vessels meeting, passing, or overtaking are never to be used except when steamers are in sight of each other and the course and position of each can be determined in the daytime by a sight of the vessel itself or by night by seeing its signal lights. In fog, mist, falling snow, or heavy rain storms, when vessels cannot see each other, fog signals only must be given.

(f) When two steam vessels are crossing, so as to involve risk of collision, the vessel which has the other on her own starboard side shall keep out of the way of the other.

(g) When a steam vessel and a sailing vessel are proceeding in such directions as to involve risk of collision, the steam vessel shall keep out of the way of the sailing vessel.

(h) Where by any of these rules, one of the two vessels is to keep out of the way, the other shall keep her course and speed.

(i) Every vessel which is directed to keep out of the way of another vessel shall, if the circumstances of the case permit, avoid crossing ahead of the other.

(j) Every steam vessel which is directed by these rules to keep out of the way of another vessel shall, on approaching her, if necessary, slacken her speed or stop or reverse.

Emphasis should be laid upon paragraphs (h) and (i). The law prescribes that when one of two vessels must keep out of the way the other must keep her course and speed. The vessel that must keep out of the way shall, if the circumstances permit, avoid crossing ahead of the other.

The following rules are quoted from the International Rules. They apply in Inland Waters:

1. Notwithstanding anything in the rules every vessel, overtaking any other, shall keep out of the way of the overtaken vessel. Every vessel coming up with another vessel from any direction more than two points abaft her beam shall be deemed an overtaking vessel. In case of doubt, the vessel is to assume herself an overtaking vessel.

2. In narrow channels every steamer shall, when it is safe and practicable, keep to the side of the fairway or mid-channel which lies on the starboard side of such vessel.

Sailing vessels underway shall keep out of the way of sailing vessels or boats fishing with lines, nets, or trawls.

DISTRESS SIGNALS

When a vessel is in distress and requires assistance from other vessels or from shore the following shall be the signals to be used or displayed by her, either together or separately:

1. *In the Daytime* a continuous sounding with any fog signal apparatus, or firing a gun.

2. *At Night* (a) Flames on the vessel as from a burning tar barrel, oil barrel, and so forth.

(b) A continuous sounding with any fog signal apparatus, or firing a gun.

PRUDENCE AND PRECAUTION

1. In obeying and construing these rules due regard shall be had to all dangers of navigation and collision, and to any special circumstances which may render a departure from these rules necessary to avoid immediate danger.

2. Nothing in the rules shall exonerate any ves-

sel, or the owner or master or crew thereof, from the consequences of any neglect to carry lights or signals, or of any neglect to keep a proper look-out, or of the neglect of any precaution which may be required by the ordinary practice of seamen, or by the special circumstances of the case.

PENALTIES

"Penalties are prescribed for the infringement of these rules by all nations that have adopted them as laws, and these penalties do not depend upon the question whether damage has or has not resulted from the infringement.

"Where damage is done, and can be shown to be the result of neglect or violation of the rules, it is held, in the absence of proof to the contrary, to be the fault of the person having charge of the deck of the vessel offending, who will be considered guilty of a misdemeanor and punishable therefor. If death ensues, he will be subject to a charge of manslaughter.

"In every case of collision, it is the duty of the person in charge of each vessel to stay by the other and render such assistance as may be practicable, provided he can do so without damage to his own ship, passengers, or crew.

"He is also required to give to the master of the

other ship the name of his own ship and of the port to which she belongs, and the ports to and from which she is bound.

"As soon as possible after the collision, he must cause an entry to be made in the log book, of the collision and of all facts connected with it."

THE END

Spotted something that isn't quite right?

Let us know and we'll fix it for future reprints.
And as a thank you for your diligence, we'll
send you a free Sothis book of your choice.

publisher@sothis.press

This page intentionally left blank

Made in United States
North Haven, CT
05 November 2026

21817326R00088